Uncovering STUDENT IDEAS About ENGINEERING and TECHNOLOGY

32 NEW Formative Assessment Probes

Uncovering STUDENT IDEAS About ENGINEERING and TECHNOLOGY

32 NEW Formative Assessment Probes

PAGE KEELEY
CARY SNEIDER
MIHIR RAVEL

Arlington, Virginia

Claire Reinburg, Director
Rachel Ledbetter, Managing Editor
Jennifer Merrill, Associate Editor
Andrea Silen, Associate Editor
Donna Yudkin, Book Acquisitions Manager

Art and Design
Will Thomas Jr., Director
Cover, Interior Design, and Illustrations
by Linda Olliver

Printing and Production
Catherine Lorrain, Director

National Science Teaching Association
1840 Wilson Blvd., Arlington, VA 22201
www.nsta.org/store
For customer service inquiries, please call 800-277-5300.

Copyright © 2020 by the National Science Teaching Association.
All rights reserved. Printed in the United States of America.
23 22 21 20 4 3 2 1

NSTA is committed to publishing material that promotes the best in inquiry-based science education. However, conditions of actual use may vary, and the safety procedures and practices described in this book are intended to serve only as a guide. Additional precautionary measures may be required. NSTA and the authors do not warrant or represent that the procedures and practices in this book meet any safety code or standard of federal, state, or local regulations. NSTA and the authors disclaim any liability for personal injury or damage to property arising out of or relating to the use of this book, including any of the recommendations, instructions, or materials contained therein.

Permissions
Book purchasers may photocopy, print, or e-mail up to five copies of an NSTA book chapter for personal use only; this does not include display or promotional use. Elementary, middle, and high school teachers may reproduce forms, sample documents, and single NSTA book chapters needed for classroom or noncommercial, professional-development use only. E-book buyers may download files to multiple personal devices but are prohibited from posting the files to third-party servers or websites, or from passing files to non-buyers. For additional permission to photocopy or use material electronically from this NSTA Press book, please contact the Copyright Clearance Center (CCC) (*www.copyright.com*; 978-750-8400). Please access *www.nsta.org/permissions* for further information about NSTA's rights and permissions policies.

Library of Congress Cataloging-in-Publication Data
Names: Keeley, Page, author. | Sneider, Cary Ivan, author. | Ravel, Mihir, 1959- author. | National Science Teaching Association, issuing body.
Title: Uncovering student ideas about engineering and technology : 32 new formative assessment probes / by Page Keeley, Cary Sneider, Mihir Ravel.
Description: Arlington, VA : National Science Teaching Association, 2020. | Includes bibliographical references and index. | English and Spanish.
Identifiers: LCCN 2019054081 (print) | LCCN 2019054082 (ebook) | ISBN 9781681403113 (paperback) | ISBN 9781681403120 (pdf)
Subjects: LCSH: Engineering--Study and teaching (Secondary)--Evaluation. | Engineering--Ability testing. | Technology--Study and teaching (Secondary)--Evaluation. | Technology--Ability testing.
Classification: LCC T65 .K34 2020 (print) | LCC T65 (ebook) | DDC 620.0071/2--dc23
LC record available at *https://lccn.loc.gov/2019054081*
LC ebook record available at *https://lccn.loc.gov/2019054082*

Contents

Section 3: Defining Problems

Section 4: Designing and Testing Solutions

Foreword

Classroom formative assessment is the most powerful form of assessment that teachers have at their disposal to elicit and analyze evidence of student thinking and, consequently, to use this evidence to adjust learning strategies accordingly. When used properly, formative assessment provides the teacher with a constant source of information that can be used during the course of and at the point of instruction. Similar to a GPS device, formative assessment is a means to keep the learner "on the path" by using student feedback as information to guide and adjust instruction. The probes detailed within *Uncovering Student Ideas About Engineering and Technology: 32 New Formative Assessment Probes* provide opportunities for students to engage in self-assessment and feedback from their peers in the areas of engineering and technology.

This book is especially timely, since 43 states and the District of Columbia have now adopted or adapted science standards based on either the *Next Generation Science Standards* (*NGSS*; NGSS Lead States 2013) or *A Framework for K–12 Science Education: Practices, Crosscutting Concepts, and Core Ideas* (the *Framework*; NRC 2012). These new state standards, which provide guidance for the development of curriculum, instruction, and assessment, call for *all students,* "over multiple years of school, [to] actively engage in scientific and engineering practices and apply crosscutting concepts to deepen their understanding of the core ideas," both in the traditional disciplines of science and in the field of engineering (NRC 2012, p. 10).

Uncovering Student Ideas About Engineering and Technology supports the vision of the *NGSS* and the *Framework* by providing educators with a variety of research-based formative assessment probes to uncover their students' prior knowledge and misconceptions in the areas of engineering and technology. This book not only offers tools for teachers to use to uncover their students' thinking, but also provides a foundation to support the importance of engineering and technology in the development of student problem-solving skills and innovative application of science concepts.

The authors of this book represent a "perfect storm" of expertise. Page Keeley is a prolific writer and researcher in the area of science formative assessment. Cary Sneider was a member of the *Framework* and *NGSS* writing teams and has worked extensively with teachers nationwide to bring engineering and design into the classroom. As a distinguished engineer, technologist, and university educator, Mihir Ravel affords his expertise through the creation of authentic, problem-based scenarios and situations addressed through the probes. The product of the collaboration of these talented experts provides the readers of this book with a practitioner-friendly guide to infusing engineering and technology into classrooms through research-based formative assessment prompts and probes.

I am honored to be asked to write the foreword for this book. Supporting educators in the implementation of three-dimensional science and engineering standards is mission critical. Teachers are the key to the positive change we seek in preparing our students to become a STEM literate citizenry. Toward that end, *Understanding Student Ideas About Engineering and Technology* goes far in supporting educators through its teacher-centered approach to engaging students and soliciting evidence of learning within the domains of engineering and technology.

—Peter J. McLaren
Executive Director
Next Gen Education, LLC

References

National Research Council (NRC). 2012. *A framework for K–12 science education: Practices, crosscutting concepts, and core ideas.* Washington, DC: National Academies Press.

NGSS Lead States. 2013. *Next Generation Science Standards: For states, by states.* Washington, DC: National Academies Press. *www.nextgenscience.org/next-generation-science-standards.*

Preface

This is the 12th book in the *Uncovering Student Ideas in Science* series. Like the other books in this series, this book provides a collection of unique questions, called formative assessment probes, that are purposefully designed to reveal preconceptions students bring to their learning as well as to identify misunderstandings students develop during instruction that may go unnoticed by the teacher. Each probe is carefully researched to develop distracters that mirror commonly held ideas students have about the key ideas or practices. The probes are not grade-specific. They are designed to be used across multiple grade spans as well as with adults for professional learning or preservice education, especially since alternative ideas that go unchallenged often follow students from one grade to the next, right into adulthood.

Engineering and Technology Probes

This book expands on the other 11 books in the *Uncovering Student Ideas* series, whose scope has been the traditional science disciplines, to help teach the emerging STEM areas of Engineering and Technology as exciting complements to Science and Mathematics. The focus is on the disciplinary content of engineering and technology, the use of engineering practices, and connections to crosscutting concepts that support an understanding of engineering and technology. As we discuss later in the book's Introduction, it is both important and empowering that the great majority of skills that students develop through engineering activities are the same as those they develop in science.

Before using the probes, please read the Introduction (pp. 1–6). The Introduction provides information on why a K–12 understanding of engineering and technology is important today not only for producing future engineers, but also for building lifelong creative and systematic problem-solving skills for students in all career paths. It describes how standards and curriculum have changed to reflect this importance. It clarifies the differences between engineering and technology and how they are inextricably linked. It provides information on how to use the probes formatively. Finally, it provides suggestions for additional resources to expand your knowledge of formative assessment as well as engineering and technology.

Four Sections of Probes

Following the Introduction is the collection of 32 formative assessment probes. These probes are organized into four sections: Section 1: What Is Technology (seven probes); Section 2: What Is Engineering? (nine probes); Section 3: Defining Problems (seven probes); and Section 4: Designing and Testing Solutions (nine probes). Each section includes a matrix that lists related key ideas and suggested grade levels for each probe. Following the matrix is a short description of teaching and learning considerations that provide additional information for refining curriculum and instruction.

Two Versions of Each Probe

There are two versions of each probe included in this book. The first is the English language student page. On the back side of the English language student page is a Spanish language version. This version can be used with English language learners or with students in Spanish language immersion programs.

Preface

Teacher Notes

Each of the 32 formative assessment probes in this book includes detailed background information for teachers. The Teacher Notes are a vital component of this book and should always be read before using a probe. The features of the Teacher Notes that accompany each probe are as follows:

Purpose

"Deciding what to assess is not as simple as it might appear. Existing guidelines for assessment design emphasize that the process should begin with a statement of the purpose for the assessment and a definition of the content domain to be measured" (Pellegrino, Chudowsky, and Glaser 2001, p. 178). This section describes the purpose of the probe—that is, what you will learn about your students' ideas as you use the probe. It begins by describing the overarching concept the probe elicits, followed by the specific idea or practice that makes up the learning target. Before choosing a probe, it is important to understand what the probe is intended to reveal about students' thinking. Taking time to read the Purpose section will help you decide if the probe will provide the information you need to plan responsive instruction and attend to students' thinking.

Type of Probe

This section describes the format used to develop the probe. All probes in the *Uncovering Student Ideas* series are two-tiered—meaning they consist of two parts. The first part is a selected answer choice and the second part involves constructing an explanation for the selected answer choice. Similar to the crosscutting concept of structure and function, in which structure often determines function, the format of a probe is related to how a probe is used. The book uses the following probe types:

- *Friendly Talk Probe:* This format uses the context of a group of friends having a conversation. Answer choices are the statements each friend makes. The probe models the importance of sharing ideas through talk and shows how people often have very different ideas.
- *Justified List Probe:* In this format, students select answer choices from a list of examples and non-examples. It shows whether students can transfer what they know or have learned to other examples or contexts and whether they can develop generalizations.
- *Opposing Views Probe:* In this format, two people have opposite or very different ideas. Selecting who to agree with involves carefully examining each statement or argument.
- *Follow the Dialogue Probe:* This format is similar to a friendly talk probe, except students follow a back-and-forth conversation in language typical of ways students converse with others.
- *Always, Sometimes, Never Probe:* This format requires students to evaluate statements to decide if they are always true or apply, sometimes true or apply, or never true or apply and then justify their answer with evidence. Selecting *sometimes* provides an opportunity to consider exceptions.
- *Draw a Picture Probe:* Unlike the format of the other probe types, students do not select a response in this probe. Instead, students draw a picture, which provides insight into their conceptual model or ways of perceiving an object, process, or situation.
- *Sequencing Probe:* This format involves students putting statements, procedures, steps, or ideas into a logical sequence.
- *Quantifying Probe:* This format involves identifying how many examples of a concept, procedure, or practice are in a given scenario.
- *Comparison Chart Probe:* This format presents students with data used to make

comparisons between the different categories of information in the chart.

To learn more about each of these probe types as well as formative assessment classroom techniques (FACTs) that can be used with these formats, see *Science Formative Assessment, Volume 1* and *Volume 2* (Keeley 2016; Keeley 2015), both available through NSTA Press.

Related Key Ideas

Each probe is designed to target one or more related key ideas that develop across multiple grade levels. A key idea represents an important aspect of understanding engineering and technology.

Explanation

The *best* answer choice is provided in this section. *Best answer* is used rather than *correct* or *right answer* because the probes are not used to pass judgment on whether students are "right or wrong," nor are they intended to be graded. Instead, they are used to encourage students to reveal their *best thinking so far* without the worry of being "wrong." Sometimes there is no single "right" answer because the probe may uncover different ways of thinking that support an alternative answer choice. In many ways, the "best answer" mirrors the nature of engineering as engineers initially share their best thinking about a design or problem situation and modify their ideas and designs as they gather more information.

A brief content explanation is provided to help teachers understand the engineering and technology ideas and practices that underlie the probe and clarify misunderstandings students (and teachers) may have related to the content. The explanations are brief and not meant to give detailed engineering and technology knowledge. They are provided to support teachers' basic knowledge of engineering and technology. Teachers with limited coursework or professional development in technology and engineering design or who are new to teaching engineering should build on these probes to expand their content knowledge. The explanations are carefully written to avoid highly technical language and complex descriptions so that a teacher does not have to specialize in engineering to understand the explanation. At the same time, the challenge is to not oversimplify the engineering concepts, key ideas, and practices. The probe explanations are carefully constructed to provide the concise information a teacher would need to understand and respond to their students' thinking.

Administering the Probe

Intended grade levels for using the probe and suggestions, including modifications, for administering the probe to students are provided. Unlike summative assessments, the probes are not specific to a single grade. They are designed to be used across grade spans even if a key idea was previously taught. Probes help teachers check for understanding of precursor ideas before introducing new ideas. They also activate student thinking by connecting their new learning to prior knowledge as well as engage students in discussions in which previous and new ideas are shared.

Connections to the Three Dimensions (NRC 2012; NGSS Lead States 2013)

A Framework for K–12 Science Education: Practices, Crosscutting Concepts, and Core Ideas (the *Framework*; NRC 2012) is the primary source document, which has informed the development of many recent state standards, including the *Next Generation Science Standards* (*NGSS*; NGSS Lead States 2013), and will continue to inform the development of most states' standards as their standards come up for revision, regardless of whether those states adopt the *NGSS*. This section lists the general disciplinary core ideas (DCIs), science and engineering practices (SEPs), and crosscutting concepts (CCCs) from the *Framework* and *NGSS* that are related to the probe.

Because the probes are not designed to be summative assessments, this section is not considered an alignment, but rather identifies ideas, practices, and concepts that are related in some way to the probe. Additional ways to support the use of the DCIs, SEPs, and CCCs are included in the Suggestions for Instruction and Assessment section.

Related Research

Each probe is informed by research when available. Research on K–12 students' ideas about engineering and technology is relatively new; therefore, there are fewer studies cited in this section compared with other books in the *Uncovering Student Ideas* series. However, consider using these probes to do your own classroom research on commonly held ideas about engineering and technology, and sharing your results with colleagues through presentations or articles in journals published by NSTA, the International Technology and Engineering Educators Association (ITEEA), and other STEM organizations.

One research article frequently cited in this book is "The Informed Design Teaching and Learning Matrix" (Crismond and Adams 2012). This meta-literature review connects research findings on how people design with what K–16 teachers need to understand and do to build student capability in engineering design and support learning through engineering design activities.

Although your students may have different backgrounds, experiences, and contexts for learning, the descriptions from the research can help you better understand the intent of each probe and the kinds of thinking your students are likely to reveal when they respond to a probe. The research also helps you understand why the distracters are written a certain way, as they are often intended to mirror research findings. As you use the probes, you are encouraged to seek new and additional published research.

Suggestions for Instruction and Assessment

Uncovering and examining the ideas students bring to their learning is considered diagnostic assessment. Diagnostic assessment becomes formative assessment when the teacher uses the assessment data in a feedback loop to make decisions about instruction that will move students toward the intended learning target. Thus, for the probe to be used formatively, a teacher needs to think about how to choose or modify a lesson or activity to best address the ideas students bring to their learning or the misunderstandings that might surface or develop during the learning process. A probe may also reveal whether students understand a key idea or use of an engineering practice, which can help the teacher move forward with planned instruction.

As you carefully analyze your students' responses, the most important next step is to make an instructional decision that would work best in your particular context. This includes considering the learning goal, your students' ideas, the materials you have available, and the diverse learners you have in your classroom.

The suggestions provided in this section have been gathered from the wisdom of teachers, the knowledge base on effective teaching, research on specific strategies used to address commonly held ideas and conceptual difficulties, and the experiences of the authors. These suggestions are not lesson plans, but rather brief recommendations that may help you plan or modify your curriculum or instruction to help students move toward learning the important ideas, concepts, and practices related to engineering and technology. It may be as simple as realizing that you need to provide a relevant, familiar problem-solving context, or there may be a specific strategy, resource, or activity that you could use with your students. Learning is a complex process and most likely no single suggestion will help all students learn. But that is what formative assessment encourages—thinking carefully about the instructional strategies, resources, and experiences needed to move students' learning forward. As you become more familiar with the ideas your students have and the multifaceted factors that may have contributed to their misunderstandings, you will identify additional strategies that you can use to teach for understanding.

References

The final section of the Teacher Notes is the list of references. References are provided for the publications cited in the Teacher Notes.

References

Crismond, D., and R. Adams. 2012. The informed design teaching and learning matrix. *Journal of Engineering Education* 101 (4): 738–797.

Keeley, P. 2015. *Science formative assessment, volume 2: 50 more strategies for linking assessment, instruction, and learning.* Thousand Oaks, CA: Corwin Press.

Keeley, P. 2016. *Science formative assessment, volume 1: 75 practical strategies for linking assessment, instruction, and learning.* 2nd ed. Thousand Oaks, CA: Corwin Press.

National Research Council (NRC). 2012. *A framework for K–12 science education: Practices, crosscutting concepts, and core ideas.* Washington, DC: National Academies Press.

NGSS Lead States. 2013. *Next Generation Science Standards: For states, by states.* Washington, DC: National Academies Press. *www.nextgenscience.org/next-generation-science-standards.*

Pellegrino, J., N. Chudowsky, and R. Glaser. 2001. *Knowing what students know: The science and design of educational assessment.* Washington, DC: National Academies Press.

Acknowledgments

We deeply appreciate the many educators whose passion and skill bring technology and engineering education to their students every day. This book is for them, and we wish to acknowledge the rewarding discussions, collaborations, and insights that we have had over the decades with many STEM educators and practitioners across the globe:

Dax Balzer
Jyotsna Bapat
Steve Barbado
Jo Barendse
Scott Bausback
John Belcher
Harald Berndt
William Berozzi
Jairo Botero
Laura Bottomley
Johanna Bunn
Barry Burke
Rodger Bybee
Beth Cady
Arthur Camins
Elizabeth Carter
Joseph Cavallaro
Mark Chang
Per Christiansen
David Coronado
David Crismond
Phil Crosby
Christine Cunningham
Martha Cyr
Alan Deale
Charlotte Denis
Tamara DePue
Rod Dougherty
Bill Dugger
Donald Duncan
Arthur Eisenkraft
Caitlin Everett
Brian Fain
Richie Faubert
Harold Foecke
Jake Foster
Andrew Fraknoi
Maurice Frazier
Dennis Freeman
Alan Friedman
Jonathan Frostad
Susan German
Tony Gordon
Alan Gould
Gabriel Grinder
Mike Hacker
Hermann Haus
Robert Heath
Luiza Holtzberg
Susan Holveck
James Hook
Tanner Huffman
Fadia Hussein
David Kaeli
Linda Katehi
Linda Kekelis
Janet Kolodner
Nancy Lapotin
Lori Lancaster
Chris Lee
Ben Linder
Robert Maybury
Mark McDermott
Beth McGrath
Joel McKee
Peter McLaren
Laura Lee McLeod
Ioannis Miaoulis
Ben Mihelic
Paige Miller
Brad Minch
Rajesh Mishra
Tamar More
Michael Morrow
Johnny Moye
Bernadine Okoro
Chetan Parikh
Carlo Parravano
Elizabeth Parry
Greg Pearson
Branimir Pejčinović
Jim Popham
S. S. Prabhu
Scott Prahl
Gill Pratt
Alyson Prior
Stephen Pruitt
Lee Pulis
Senay Purzer
Helen Quinn
N. J. Rao
Ted Rappaport
Neelam Ravel
Diane Reid
Dan Robinette
Ajay Rundell
S. N. Sadagopan
Steve Scannell
Dennis Schatz
Gunnar Scherner
Bruce Shafer
Jomae Sica

Acknowledgments

Bill Siebert
Charlie Sodini
Mark Somerville
Yvonne Spicer
Kendall Starkweather
Ray Stata
Lynn Stein
Jon Stolk
Brian Storey
Johannes Strobel
Melissa Tollinger
Julie Trisel
Jim Truchard
Joyce Tugel
J. Kim Vandiver
Jennifer Wells
Ryan Windle
Tong Zhang

We are also grateful to our publisher, NSTA Press, for adding this book to NSTA's growing collection of engineering resources. We especially would like to thank Claire Reinburg, Rachel Ledbetter, Linda Olliver, and Kate Hall for their superb skill and support in transforming this work from manuscript to publication. We also give a special thank you to two outstanding teachers for checking and refining the Spanish translations: Jose Rivas from Lennox Academy in Inglewood, California, and Adelina Martinez from Hatch Valley Middle School in Hatch, New Mexico.

Dedication

This book is dedicated to the memory of two individuals whose vision and cooperative spirit have helped bring together the worlds of K–12 technology and engineering education with the world of science education:

William E. Dugger, Jr., who led the development of the *Standards for Technological Literacy,* the first set of educational standards for technology and engineering; and

Alan J. Friedman, who helped advance the integration of science inquiry and engineering design in schools and beyond the school day through a lifetime of accomplishment.

About the Authors

Page Keeley is the primary author of the *Uncovering Student Ideas in Science* series. Her assessment probes and FACTs (formative assessment classroom techniques) are widely used by K–12 teachers, university professors, and professional development and science specialists throughout the United States and internationally. Page is "retired" from the Maine Mathematics and Science Alliance (MMSA) where she had been the senior science program director since 1996, directing projects in the areas of instructional leadership, coaching and mentoring, linking standards and research, and science and literacy. She has been a principal investigator and project director of three National Science Foundation (NSF)–funded projects: The Northern New England Co-Mentoring Network (NNECN), Curriculum Topic Study (CTS), and Phenomena and Representations for Instruction of Science in Middle School (PRISMS). Today she works as an independent consultant, speaker, and author providing professional development to school districts and organizations in the areas of formative assessment, understanding student thinking, teaching science for conceptual understanding, and designing effective instruction.

Page is a prolific author of 22 national bestselling and award-winning books in science and mathematics education. Several of her books have received national distinguished awards in educational publishing. She has authored numerous journal articles and contributed to several book chapters. She also develops formative assessment probes for McGraw-Hill's middle and elementary school science programs.

Prior to joining MMSA in 1996, Page taught middle and high school science for 15 years. At that time she was an active teacher leader at the state and national levels, serving as president of the Maine Science Teachers Association and National Science Teaching Association (NSTA) District II Director. She received the Presidential Award for Excellence in Secondary Science Teaching in 1992, the Milken National Distinguished Educator Award in 1993, and the AT&T Maine Governor's Fellowship in 1994. Since leaving the classroom in 1996, her work in leadership and professional development has been nationally recognized. In 2008, she was elected the 63rd president of NSTA. In 2009, she received the National Staff Development Council's (now Learning Forward) Susan Loucks-Horsley Award for Leadership in Science and Mathematics Professional Development. In 2013, she received the Outstanding Leadership in Science Education award from the National Science Education Leadership Association (NSELA), and she received the NSTA Distinguished Service to Science Education Award in 2018. She has served as an adjunct instructor at the University of Maine, was a Cohort 1 Fellow in the National Academy for Science and Mathematics Education Leadership, was a science literacy leader for the AAAS/Project 2061 Professional Development Program, and has served on several national advisory boards.

About the Authors

She has led science/STEM education delegations to South Africa (2009), China (2010), India (2012), Cuba (2014), Iceland (2017), Panama (2018), and Costa Rica (2019).

Prior to entering the teaching profession, Page was a research assistant for immunogeneticist Dr. Leonard Shultz at the Jackson Laboratory of Mammalian Genetics in Bar Harbor, Maine. She received her BS in life sciences and pre-veterinary studies from the University of New Hampshire and her MEd in science education from the University of Maine. In her spare time she enjoys travel, reading, fiber art, and photography, and also dabbles in modernist cooking and culinary art. A Maine resident for almost 40 years, Page and her husband now divide their time between homes in Fort Myers, Florida, and Wickford, Rhode Island.

You can contact Page through her websites at *www.uncoveringstudentideas.org* and *www.curriculumtopicstudy2.org* or via e-mail at pagekeeley@gmail.com. You can follow her on Twitter at @CTSKeeley or on Facebook through her Uncovering Student Ideas in Science and Mathematics page.

Cary Sneider is a visiting scholar at Portland State University, and a consultant for the STEM Next Opportunity Fund and the Stephen D. Bechtel Jr. Foundation, both charitable foundations that support STEM education. He also continues to be active as a consulting author for Houghton Mifflin Harcourt's Science Dimensions K–8 series, and other writing and curriculum development projects.

While studying astrophysics at Harvard College in the 1960s, Cary volunteered to teach in an Upward Bound program and discovered his real calling as a science teacher. After teaching middle and high school science in Maine, California, Costa Rica, and Micronesia, he settled for nearly three decades at Lawrence Hall of Science in Berkeley, California, where he developed skills in curriculum development and teacher education.

Starting in 1997, Cary spent 10 years as vice president for programs at the Museum of Science in Boston, where he led development of a high school engineering curriculum, Engineering the Future: Science, Technology, and the Design Process. In 2007, he moved to Portland, Oregon, to take a position as associate research professor at Portland State University, where for the next decade he taught courses in research methodology to more than 80 candidates in a Master of Science Teaching (MST) program. During this period, he led a team in revising science education standards for the state of Washington; served as a consultant to the National Research Council to help create *A Framework for K–12 Science Education: Practices, Core Ideas, and Crosscutting Concepts*; led the engineering team that helped craft the *Next Generation Science Standards: For States, by States*, which were released in 2013; and from 2011 to 2019 served as a member of the National Assessment Governing Board, which sets policy for the National Assessment of Educational Progress (NAEP), also known as The Nation's Report Card.

In 1997, Cary received the Distinguished Informal Science Education award from the National Science Teaching Association (NSTA). In 2003, he was named National Associate of the National Academy of Sciences, and in 2018 he received the Robert H. Carleton Award, NSTA's highest recognition.

Over his career, Cary has directed more than 20 federal, state, and foundation grant projects. He has coauthored two books on STEM education, edited the three-volume

About the Authors

Go-To Guide to Engineering Curricula; authored *Jake and the Quake*, a work of historical fiction for middle school students; and authored several book chapters and numerous articles. He earned a BA in astronomy at Harvard College, and a California Secondary Teaching Credential, MA, and PhD in science education at the University of California, Berkeley.

Mihir Ravel is a noted technology leader in high-performance electronic systems and ultrafast scientific measurements, and a pioneer in design-centric approaches to integrated STEM education.

After a fortunate corporate research and development (R&D) career, Mihir now divides his time between public service in K–12 engineering and science education and advising entrepreneurs and innovators focused on emerging opportunities for social good. He has traveled extensively in developed and developing countries as a speaker and strategic adviser to both public and private institutions. He has been an international advisor to various federal and state agencies on STEM and design education; has served on the advisory boards for EDN magazine, the Austin Technology Council, the Massachusetts Institute of Technology (MIT) Enterprise Forum, and The Indus Entrepreneurs (TiE); and is an adviser to various early stage private ventures and nonprofit educational initiatives and foundations.

Mihir has been an invited faculty collaborating on design-centric learning methods with leading universities in Europe, Asia, and the United States, with a special emphasis on exposing students to the power of engineering design and entrepreneurship for making a better world. A highlight of his university collaborations was helping incubate the Affordable Design and Entrepreneurship (ADE) program at Olin College of Engineering, which was recently recognized alongside MIT as one of the top two global leaders in engineering education. The ADE program is a transformational educational initiative aimed at the problems of global poverty, and immerses student teams in creating social ventures using an engineering design process as a critical tool for improving the daily lives of the world's poor.

To build on his lessons learned in advancing university education, Mihir has recently been applying those experiences toward design-based approaches to K–12 STEM education. He has partnered with school districts, universities, and foundations in the Pacific Northwest to develop engineering and design–integrated STEM curricula at the high school level. He is a coauthor with Cary Sneider and others of the second edition of *Engineering the Future: Science, Technology, and the Design Process,* developed in collaboration with the Boston Museum of Science and Activate Learning, as an introduction to engineering design for *all* students, not just those interested in engineering careers. He is also an adviser to Houghton Mifflin Harcourt on K–12 science.

Complementing his university experience, Mihir's STEM education work builds on lessons learned from three decades of leading high-tech R&D organizations in a range of technologies spanning ultrafast electronics, optical and wireless communications, digital multimedia, environmental monitoring, and smart sensor networks. He has led research collaborations with leading university and government labs with support from the National Science Foundation, Department of Energy, and Department of Defense. He served as the first vice president of technology for National Instruments, a global pioneer in tools for

About the Authors

measurement, machine automation, and system design. Previously, he was a Technology Fellow and head of Strategic Technologies for Tektronix, a global leader in high performance instrumentation and electronic design. His perspective that STEM education should be a combination of both creative and structured thinking has been shaped by his early training in physics, electrical engineering, and computer science at MIT.

Introduction

What Are Formative Engineering and Technology Probes?

The subject of *Uncovering Student Ideas About Engineering and Technology: 32 New Formative Assessment Probes* highlights the biggest change in the content of K–12 science education in more than a century—that engineering be taught alongside the traditional disciplines of life, physical, and Earth and space science. Although initially it may seem like technology and engineering are two different subjects, they are actually two sides of the same coin. Technology is the designed world—everything around us that has been created by people; engineering is the process of inventing and improving technologies.

Readers who are familiar with the other 11 books in the *Uncovering Student Ideas* series will already know how to use these probes. If not, you'll catch on as soon as you try one with your students. Each probe is a conversation-starter, designed to uncover your students' pre-existing ideas. They become formative when you use the information about your students' thinking to make informed instructional decisions that will help them modify or refine their initial ideas.

Why Technology and Engineering Design Are Essential for ALL Students

There are very good reasons why technology and engineering have risen to prominence in K–12 education. In 1950, the global population was about 2.5 billion people, by 2015 it had more than tripled to 7.3 billion, and estimates project it to be about 10 billion by 2050 (United Nations Population Division, *https://population.un.org/wpp*). How will future generations meet the growing needs of this population? All students must understand that engineering and technology are powerful tools to meet our escalating needs for affordable health care and housing, clean energy, efficient transportation, nourishing food, and clean water. Just as learning how the natural world functions (science) is critical to understanding these problems, equally so is the process of solving them through engineering.

In today's modern society, in which we are all surrounded by complex technologies and expected to make technological decisions on a daily basis as consumers, workers, and citizens, it is essential for everyone to become technologically literate and to be able to apply user-centered design approaches to solving problems in their daily lives. That is why technology and engineering education are important for ALL students, not just those who will become tomorrow's engineers.

Changes in Science Education Standards

The International Technology and Engineering Educators Association (ITEEA) has been a pioneer in developing K–12 standards for ***all students*** to learn about engineering and technology. *Standards for Technological Literacy: Content for the Study of Technology* (ITEEA 2007) identified 20 standards: seven on the nature of technology and its relationship to society; six on technology and engineering abilities; and seven on modern civilization's major technological systems, including medical, agricultural, transportation, and energy systems.

Introduction

Each standard includes benchmarks for grades K–2, 3–5, 6–8, and 9–12, to guide the work of teachers and curriculum developers. Many of these ideas and capabilities have since been incorporated in *A Framework for K–12 Science Education: Practices, Crosscutting Concepts, and Core Ideas* (the *Framework*; NRC 2012) and the subsequent *Next Generation Science Standards* (*NGSS*; NGSS Lead States 2013).

At the time this book was being written, a great number of states in the United States have adopted or adapted science education standards that include engineering as a core subject at the same level as life science, physical science, and Earth and space science. In addition to having its own set of core ideas and performance expectations in the new standards, engineering as a practice is to be fully integrated with the other science disciplines. Although engineering has long been a part of science standards and curricula, in the past it has usually been seen as a way to reinforce science concepts by expecting students to apply what they learned in more traditional science classes. Also, it has been used primarily to teach topics in physics, such as force and motion, energy, and waves, and only rarely applied to other fields of science. In contrast, the vision of engineering in the *Framework* (NRC 2012) and the *NGSS* (NGSS Lead States 2013) is that students are expected to be able to apply an engineering design process to all fields of science, to understand how science and engineering drive each other forward, and to solve real-world problems by considering the ways that science, technology, and engineering interact with society and the natural world. As explained in volume 2 of the *NGSS* (NGSS Lead States 2013, p. 3):

> *The rationale for this increased emphasis on engineering and technology rests on two positions taken in the* Framework. *One position is aspirational, the other practical.*
>
> *From an aspirational standpoint, the* Framework *points out that science and engineering are needed to address major world challenges such as generating sufficient clean energy, preventing and treating diseases, maintaining supplies of food and clean water, and solving the problems of global environmental change that confront society today. These important challenges will motivate many students to continue or initiate their study of science and engineering.*
>
> *From a practical standpoint, the* Framework *notes that engineering and technology provide opportunities for students to deepen their understanding of science by applying their developing scientific knowledge to the solution of practical problems. Both positions converge on the powerful idea that by integrating technology and engineering into the science curriculum, teachers can empower their students to use what they learn in their everyday lives.*

Not surprisingly, it is taking time to integrate engineering into school curricula. Results from the *Report of the 2018 National Survey of Science and Mathematics Education* highlight this issue (Banilower et al. 2018). Among elementary teachers, most feel well prepared or very well prepared to teach life science (75%), Earth science (71%), and physical science (59%). However, only 9% feel well or very well prepared to teach engineering. And although middle school and high school science teachers are generally more confident than elementary teachers, only 6% of middle school teachers and 7% of high school science teachers are very confident in their abilities to teach engineering. Keeping in mind that the *NGSS* includes engineering as a fourth discipline, note that only 46% of high schools offer engineering courses, compared with 97% that offer biology, 94% that offer chemistry, 84% that offer physics, and 59% that offer Earth science. On the other hand, that is a big improvement since the last survey in 2012, when only 24% of high schools offered courses in engineering (Banilower et al. 2013).

This book is intended to speed and deepen the process of integrating engineering into the school curriculum by providing teachers with tools to assess their students' understanding of technology and engineering, using the method of "assessment probes" exemplified in the *Uncovering Student Ideas* series. This introduction provides a brief orientation to this set of probes by explaining the authors' perspective on the meaning of technology and engineering (and why technology and engineering are essential for all students to learn), how the probes are organized into four sections, and additional NSTA resources to extend your learning.

The Meaning of Technology and Engineering

Technology and engineering are intimately related, but they are not the same. The *Framework* describes the relationship between these terms as follows:

> *In the K–12 context, science is generally taken to mean the traditional natural sciences: physics, chemistry, biology, and (more recently) earth, space, and environmental sciences. … We use the term engineering in a very broad sense to mean any engagement in a systematic practice of design to achieve solutions to particular human problems. Likewise, we broadly use the term technology to include all types of human-made systems and processes—not in the limited sense often used in schools that equates technology with modern computational and communications devices. Technologies result when engineers apply their understanding of the natural world and of human behavior to design ways to satisfy human needs and wants. (NRC 2012, pp. 11–12)*

Since many teachers who use these probes are likely to be science teachers, it is important to point out that the great majority of skills that students develop through engineering activities are the same as those they develop in science. The eight science and engineering practices identified in the *Framework* and *NGSS* are the same, whether students are exploring the natural world or improving the designed world. The major difference is in the goal of the two activities. The aim of science is to understand the natural world, while engineering aims to solve a problem or meet a need. When engaged in solving a problem, it is essential for students to learn about people's needs and desires that require and inspire the development of new and improved technologies. Doing so requires persistence and logical thinking, coupled in equal measure with imagination and compassion for the people in need. Engineering is more than applied science. It is a creative art grounded in compassion for serving society and protecting the natural world.

Organization of This Book

The probes in this book are divided into the following four sections.

- **Section 1: What Is Technology?** Before students learn about engineering, they must recognize that they are surrounded by technologies that have been designed and improved by engineers. This section will reveal your students' understanding of the nature and purpose of technologies and how technologies change over time. Other probes in this section focus on technological processes and the critical idea of systems.

- **Section 2: What Is Engineering?** This section begins with a series of probes about students' initial ideas concerning who can become an engineer and what motivates engineers. Other probes reveal students' ideas about how engineers work together in teams, the similarities between science and engineering, and the nature of engineering—spanning the range from creative to logical, imaginative to systematic, and scientific to mathematical.
- **Section 3: Defining Problems.** Beginning with a very important foundational probe about the basics of an engineering design process (EDP), this group of probes is designed to reveal your students' understanding of the critical importance of defining a problem before beginning to solve it. Probes include the practice of defining the problem, determining who really needs the solution, whether or not the problem can be solved through engineering, and how to define the problem in terms of criteria and constraints. It ends with a probe about the value of research to take advantage of prior successes and user feedback.
- **Section 4: Designing and Testing Solutions.** Once your students have defined the problem, how do they go about designing a solution? The first part of this section includes probes intended to help you refine instruction to address your students' current thinking about the process of brainstorming new ideas, drawing ideas from nature, considering the affordability and sustainability of designs, and the process of sorting through initial ideas to choose the one that is most promising, making trade-offs, and using science and math to design a successful solution.

  Designs can only be improved and verified to work when they've been tested—testing is actually the phase of an EDP where engineers (and students!) often learn the most about a design: what's working, what isn't working, and why. The second part of this section begins with a sequence of probes concerning the different kinds of models used in comparing and developing designs, from sketches and technical drawings, to computer simulations and physical prototypes. Other probes concern the process of choosing the best solution, the pervasive role of science and math in engineering design, and iterating and optimizing the design to make it better.

Although the probes in this book do not cover every aspect of engineering and technology, the entire collection should provide sufficient scope to help you develop a fairly in-depth understanding of your students' current vision of engineering design—not only as a school subject that they are expected to learn, but also to learn the skills of systematic problem solving and teamwork needed to solve the challenges facing our world, as well as ones that may arise in their own lives.

Additional NSTA Resources

NSTA has numerous resources to support your use of the probes in this book. In addition to the NGSS@NSTA Hub (*https://ngss.nsta.org*), the following resources may also be useful.

Formative Assessment Resources

Keeley, P. 2014. *What are they thinking? Promoting elementary learning through formative assessment.* Arlington, VA: NSTA Press.

Keeley, P. 2015. *Science formative assessment, volume 2: 50 more strategies for linking assessment, instruction, and learning.* Thousand Oaks, CA: Corwin Press (a co-publication with NSTA Press).

Keeley, P. 2016. *Science formative assessment, volume 1: 75 practical strategies for linking assessment, instruction, and learning.* 2nd ed. Thousand Oaks, CA: Corwin Press (a co-publication with NSTA Press).

Keeley, P. 2019. Guest editorial: Formative assessment in the science classroom: What it is and what it is not. *Science and Children* 56 (9): 8–9.

Keeley P. *Uncovering Student Ideas in Science Series. www.nsta.org/publications/press/uncovering.aspx.* These probes can be used to assess students' ideas about disciplinary content knowledge used in an engineering problem.

Science and Children Journal (2010–present). Page Keeley writes a monthly column titled "Formative Assessment Probes: Promoting Learning Through Assessment." While written for the elementary journal, the suggestions in the column apply to K–12. There are more than 60 articles in this collection (*www.nsta.org/elementaryschool*).

To learn more about professional development support for formative assessment, visit the *Uncovering Student Ideas* website at *www.uncoveringstudentideas.org.*

Engineering and Technology Resources

Kanter, D. E., and D. P. Crismond. 2017. Core idea ETS1: Engineering design. In *Disciplinary core ideas: Reshaping teaching and learning,* ed. R. Duncan, J. Krajcik, and A. Rivet, 245–262. Arlington, VA: NSTA Press.

Keeley, P., and J. Tugel. 2019. *Science curriculum topic study: Bridging the gap between three-dimensional standards, research, and practice.* 2nd ed. Thousand Oaks, CA: Corwin Press. This book includes several curriculum topic study (CTS) guides for understanding disciplinary core ideas about engineering; engineering practices; and connections to science, technology, and society.

Schwarz, C. V., C. Passmore, and B. J. Reiser, eds. 2017. *Helping students make sense of the world using next generation science and engineering practices.* Arlington, VA: NSTA Press.

Sneider, C. 2012. Core ideas of engineering and technology. *The Science Teacher* 79 (1): 32–36.

Sneider, C. 2017. Core idea ETS2: Links among engineering, technology, science, and society. In *Disciplinary core ideas: Reshaping teaching and learning,* ed. R. Duncan, J. Krajcik, and A. Rivet, 263–277. Arlington, VA: NSTA Press.

Willard, T. 2020. *The NSTA atlas of the three dimensions.* Arlington, VA: NSTA Press. This resource includes several conceptual mapped progressions of the disciplinary core ideas and practices related to engineering and technology.

NSTA Press books on engineering. Go to the NSTA Store at *www.nsta.org/store* and enter "engineering" in the search bar. NSTA has several books on engineering and engineering design challenges in its STEM collection.

NSTA Web Seminar: Engineering Design as a Core Idea, presented by Cary Sneider. *www.youtube.com/watch?v=dTh_kTfOsDA.*

Science and Children elementary school journal articles. Go to *www.nsta.org/elementaryschool* and enter "engineering and technology" in the Journal archives search bar.

Science Scope middle school journal articles. Go to *www.nsta.org/middleschool* and enter "engineering and technology" in the Journal archives search bar.

The Science Teacher high school journal articles. Go to *www.nsta.org/highschool* and enter "engineering and technology" in the search bar.

References

Banilower, E. R., P. S. Smith, K. A. Malzahn, C. L. Plumley, E. M. Gordon, and M. L. Hayes. 2018. *Report of the 2018 National Survey of Science and Mathematics Education (NSSME+).* Chapel Hill, NC: Horizon Research, Inc.

Banilower, E. R., P. S. Smith, I. R. Weiss, K. A. Malzahn, K. M. Campbell, and A. M. Weis. 2013. *Report of the 2012 National Survey of*

Science and Mathematics Education. Chapel Hill, NC: Horizon Research, Inc.

International Technology and Engineering Educators Association (ITEEA). 2007. *Standards for Technological Literacy: Content for the study of technology.* 3rd ed. Reston, VA: ITEEA. *www.iteea.org/File.aspx?id=67767.*

National Research Council (NRC). 2012. *A framework for K–12 science education: Practices, crosscutting concepts, and core ideas.* Washington, DC: National Academies Press.

NGSS Lead States. 2013. *Next Generation Science Standards: For states, by states.* Washington, DC: National Academies Press. *www.nextgenscience.org/next-generation-science-standards.*

Section 1
What Is Technology?

Key Ideas Matrix for Probes #1–#7

PROBES	#1 Surrounded by Technologies	#2 Is It a Technology?	#3 What's the Purpose of Technology?	#4 How Do Technologies Change?	#5 Block Diagrams	#6 Technology, System, or Both?	#7 Systems Within Systems
GRADE-LEVEL USE →	**3–12**	**3–12**	**3–12**	**3–12**	**3–12**	**3–12**	**6–12**
RELATED KEY IDEAS ↓							
Technology is any modification of the natural world to fulfill human needs or desires.		X	X	X			
Technology includes all types of human-made systems and processes, not just modern electrical devices.	X	X	X				
We live in a world in which we are surrounded by technologies.	X	X					
Technology can be used to preserve and improve the environment.			X				
People's needs and wants change over time, as do their demands for new and improved technologies.				X			
Technologies consist of products, processes, and systems.					X	X	X
A block diagram is a means for showing a technological system or process.					X		
A system is composed of two or more parts, and its function differs from that of the individual parts.						X	X
Technological systems are nested within larger systems.							X

Teaching and Learning Considerations

Before students learn about engineering, it is important for them to recognize that they are surrounded by technologies. Even the apparently "natural" environment consisting of tree-lined streets and gardens has been planted by people, using a variety of cultivation technologies. This section will reveal your students' ideas about the nature and purpose of technology, how technologies change over time, and technological processes and systems.

The probes in this section are based on a number of research studies showing that few people understand the broad definition of technology put forward in *A Framework for K–12 Science Education* (the *Framework*; NRC 2012) and *Next Generation Science Standards* (*NGSS*; NGSS Lead States 2013). Used as a springboard for discussion, these probes can help your students broaden their understanding of the technologies that they depend on in their daily lives, and that most people take for granted.

Key ideas introduced through these probes include the definition and purpose of technology; how technologies change over time; and the nature of technology as products, processes, and systems.

An extensive overview of the nature of technology; key technological concepts; and the effects of technology on the environment, society, and human history for various K–12 grade levels is available in the *Standards for Technological Literacy* (ITEEA 2007) on pages 21–87.

References

International Technology and Engineering Educators Association (ITEEA). 2007. *Standards for Technological Literacy: Content for the study of technology.* 3rd ed. Reston, VA: ITEEA. *www.iteea.org/File.aspx?id=67767.*

National Research Council (NRC). 2012. *A framework for K–12 science education: Practices, crosscutting concepts, and core ideas.* Washington, DC: National Academies Press.

NGSS Lead States. 2013. *Next Generation Science Standards: For states, by states.* Washington, DC: National Academies Press. *www.nextgenscience.org/next-generation-science-standards.*

Surrounded by Technologies

Reggie: It looks like we have only cold sandwiches today—I heard the cafeteria had a problem. It sure would be simpler if there was no more technology. Then, things would just work right all the time.

Ebony: Are you joking? If there was no more technology, the lights would go out too, so we couldn't see what we were eating.

Tishon: That's not all. We wouldn't even be eating these cold sandwiches, because bread is a technology.

Reggie and Ebony: Huh? What do you mean by that, Tishon?

Do you agree with Tishon that "bread is a technology"?

___ Yes, I agree with Tishon. ___ No, I disagree with Tishon.

Explain your thinking. How did you decide whether bread is a technology?

___

___

___

___

___

Rodeado de Tecnología

Reggie: Parece que hoy solo tendremos sándwiches fríos. Escuché que la cafetería tenía un problema. Seguro que sería más simple si no hubiera más tecnología, entonces las cosas funcionarían correctamente todo el tiempo.

Ebony: ¿Estás bromeando? Si no hubiera más tecnología, las luces también se apagarían, por lo que no podríamos ver lo que estábamos comiendo.

Tishon: Eso no es todo. No estaríamos comiendo estos sándwiches fríos, porque el pan es una tecnología.

Reggie y Ebony: ¿Eh? ¿Qué quieres decir con ese, Tishon?

¿Estás de acuerdo con Tishon en que «el pan es tecnología»?

___ Sí, estoy de acuerdo con Tishon. ___ No, no estoy de acuerdo con Tishon.

Explica lo que piensas. ¿Cómo decidiste si el pan es una tecnología?

____

____

____

____

____

Surrounded by Technologies

Teacher Notes

Purpose

The purpose of this assessment probe is to elicit students' understanding of the word *technology*. The probe is designed to find out if students are aware that they live in a world composed almost entirely of technological objects, systems, and processes.

Type of Probe

Follow the dialogue

Related Key Ideas

- Technology includes all types of human-made systems and processes, not just modern electrical devices.
- We live in a world in which we are surrounded by technologies.

Explanation

The best answer is "Yes, I agree with Tishon." A broad definition of *technology* is all the products, processes, and systems that people develop to improve the human-made world and preserve the natural environment. Technology includes everything that people make and the processes people use to make things. Bread starts with materials from nature—wheat, yeast, salt, and other natural ingredients (although some strains of wheat and yeast are human engineered). Bakers process and bake these different ingredients into the bread products needed by society.

Administering the Probe

This probe is best used with students in grades 3–12. The probe can be extended by asking students, "What if all the technology in the room were to disappear? What would the three friends see when they looked around?"

Connection to the Three Dimensions (NRC 2012; NGSS Lead States 2013)

- DCI: ETS2.B. Influence of Science, Engineering, and Technology on Society and the Natural World

Related Research

- A Gallup poll in 2002 found that "the American public is virtually unanimous in regarding the development of technological

literacy as an important goal for people at all levels. … [and] there is near total consensus in the public sampled that schools should include the study of technology in the curriculum." Nonetheless, "many Americans view technology narrowly as mostly being computers and the Internet" (Rose and Dugger 2002, p. 1).

- In a learning study, Lachapelle et al. (2013, p. 4) found that on a pre-assessment of students' understanding of technology, "students generally described technology as something powered by electricity or as the electricity that powers such devices. No conceptions other than 'human made' and 'other' showed up in more than 6% of student responses." After a unit of study in the Engineering is Elementary curriculum, students were much more likely to express the more normative conception that technology is something that is human made to solve a problem or meet a need, and were much less likely to describe technology as only something electrical.
- Roth (2001) taught a four-month technology-centered curriculum in a split sixth and seventh grade class as a means for students to learn about simple machines, energy, and forces. The researcher concluded that the design activities, and in particular opportunities for students to discuss their ideas, gave rise to a deeper understanding of the core concepts.

Suggestions for Instruction and Assessment

- Ask students to look at all the objects around them and decide which are technologies. Explain that sometimes people use the word *technology* to just refer to computers, smartphones, or other things that run on a battery, or can be plugged in. Those things are *modern* technologies. A broad definition of *technology* is all the products, processes, and systems that people develop to improve the human-made world and preserve the natural environment. Ask students to look at the objects again with that definition in mind and see if they have changed their minds about any of the objects.
- Ask what else the students see that is "technology." Eventually, the students should realize that the building they are in and even their clothes are technologies. Ask what it would look like if all the technologies around them were to disappear. (If all the technologies were to disappear, they would be standing in a forest or perhaps a meadow or lake.)
- It is important to help students understand the definitions of science, technology, and engineering, and how they are related:
    - **Science:** A systematic process of asking and answering questions about the natural or designed world, leading to greater understanding and predictive power.
    - **Engineering:** A systematic process for defining problems and developing solutions to improve the human-made world while preserving the natural environment for future generations.
    - **Technology:** All the products, processes, and systems that people develop to improve the human-made world and preserve the natural environment.
    - **Science, technology, and engineering are closely related** because scientists and engineers work together to create and improve technologies to meet people's needs.
- Have students suggest their own examples of technology that might not be commonly thought of as technology.

References

Lachapelle, C. P., J. D. Hertel, J. Jocz, and C. M. Cunningham. 2013. Measuring students' naïve conceptions about technology. Paper presented at the NARST Conference, Rio Grande, Puerto Rico.

National Research Council (NRC). 2012. *A framework for K–12 science education: Practices, crosscutting concepts, and core ideas.* Washington, DC: National Academies Press.

NGSS Lead States. 2013. *Next Generation Science Standards: For states, by states.* Washington, DC: National Academies Press. *www.nextgenscience.org/next-generation-science-standards.*

Rose, L. C., and W. E. Dugger, Jr. 2002. ITEA/Gallup poll reveals what Americans think about technology. *The Technology Teacher* 61 (6): 1–8.

Roth, W.-M. 2001. Learning science through technological design. *Journal of Research in Science Teaching* 38 (7): 768–790.

Is It a Technology?

Circle all the examples of technology.

 Pencil	 Airplane	 Strawberry	 Horse
 Frog	 Taco Shell	 Soccer Ball	 Pie
 Flower	 Lamp	 Bear	 Lightning
 Eggs	 Cell Phone	 Candle	 Cup
 Volcano	 Bird	 Balloons	 Hat

Source: Concept adapted from Cunningham (2018).

Explain your thinking. What "rule" or reasoning did you use to decide which things are a technology?

___

___

___

___

¿Es una Tecnología?

¿Qué cosas son ejemplos de tecnología?

 Lápiz	 Avión	 Fresa	 Caballo
 Rana	 Taco Dorado	 Fútbol	 Tarta
 Flor	 Lámpara	 Oso	 Relámpago
 Huevos	 Teléfono Móvil	 Vela	 Taza
 Volcán	 Pájaro	 Globos	 Sombrero

Origen: Concepto adaptado de Cunningham (2018).

Explica lo que piensas. Describe la «regla» o racionamiento que usastes para decir qué cosas son una tecnológia.

___

___

___

___

Is It a Technology?

Teacher Notes

Purpose

The purpose of this assessment probe is to elicit students' ideas about what is and what is not technology. The probe is designed to find out if students have a broad understanding of technology as defined in the *Framework* (NRC 2012).

Type of Probe

Justified list

Related Key Ideas

- Technology is any modification of the natural world to fulfill human needs or desires.
- Technology includes all types of human-made systems and processes, not just modern electrical devices.
- We live in a world in which we are surrounded by technologies.

Explanation

The best answer is pencil, airplane, taco shell, soccer ball, pie, lamp, candle, cell phone, balloons, cup, and hat. *Technology* is defined in the *Framework* as "any modification of the natural world made to fulfill human needs or desires" (NRC 2012, p. 202). Given this definition, the remaining objects in the illustration are natural objects, and thus are not considered to be technologies. (However, arguments could be made that some of the objects, such as the strawberry and horse, are not entirely "natural" because they have been bred for particular characteristics.)

Administering the Probe

This probe is best used with students in grades 3–12. It can be used with interactive formative assessment strategies such as card sorts (Keeley 2016). First, check to make sure students are familiar with the objects on the list. You can extend the probe by adding additional natural and human-designed objects.

Connection to the Three Dimensions (NRC 2012; NGSS Lead States 2013)

- DCI: ETS2.B. Influence of Science, Engineering, and Technology on Society and the Natural World

Related Research

- This probe is adapted from instruments used in a series of research studies (e.g., Jocz and Lachapelle 2012; Lachapelle et al. 2013) to evaluate the effectiveness of an elementary curriculum, Engineering is Elementary (EiE). The researchers scored the students' responses as "correct" if they circled examples of technologies and did not circle natural objects. The findings of the research studies were summarized in the following text and figure:

  Using these instruments, we found that our engineering curriculum has a dramatic, significant impact on broadening students' understandings of technology. A number of studies using control groups reinforced the fact that students gain a more accurate and nuanced understanding of technology after engaging in engineering. As [the figure] shows, after completing an EiE unit, students are much more likely to indicate that commonplace, simple technologies, such as brooms, baskets, and bicycles, are technologies. In response to the open-ended question 'How do you know if something is technology?' students are more likely to answer that technologies are human-made, and that technologies are designed to solve problems. (Cunningham 2018, p. 127)

Frequency of correct answers before and after instruction

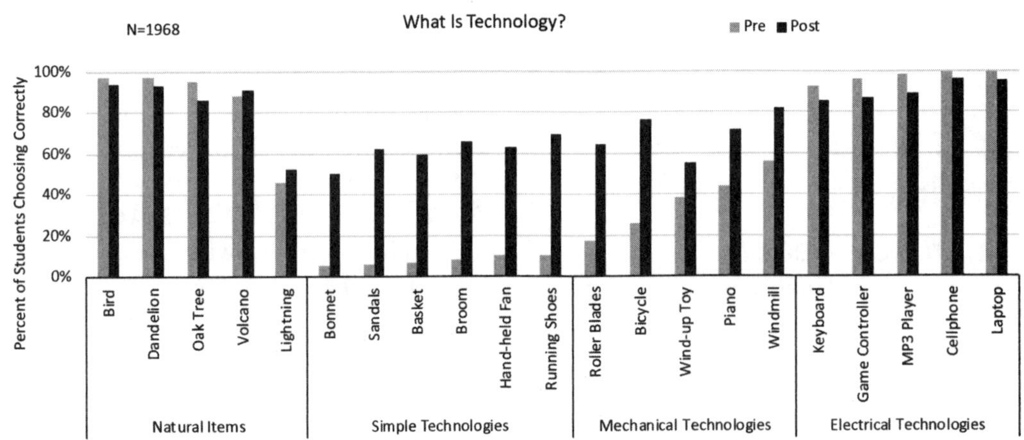

Source: Adapted from Cunningham 2018, Figure 7.1, p. 127.

Suggestions for Instruction and Assessment

- Provide students with a broad definition of *technology*: any modification of the natural world to fulfill human needs or desires. Given that definition, would they change their minds about any of the objects?
- This probe can be used as a card sort. Print each item on a card and distribute a set of cards to each pair or small group of students. Students sort the cards into three columns: Examples of Technology, Not Examples of Technology, Unsure, or We Do Not All Agree. Students must discuss each card before placing it in a column. When finished, students come up with their "rule" or reason for selecting which cards are examples of technology. Results are shared with the class.
- Have students come up with a label for the set of objects that are not examples of technology. Listen to see if they recognize that these are natural objects.
- Have the class vote on each object in the probe to see if they think it is a technology or a natural object. It's best not to give your opinion when students don't agree, so they will continue to question each other about their understanding of technology.
- Take the class outdoors and challenge them to find a technology. Then have them find a natural object. Ask them the following: Which was more difficult to find, or took longer? Why do you think that's the case?
- A method to help students understand the concept of technology is to use the Frayer model for learning vocabulary (Keeley 2016). Students write the word *technology* in the middle of the page, and divide the rest of the page into four quadrants. In one quadrant they list the definition of technology. In the second they list characteristics of technology. In the third they list technology examples. In the fourth quadrant they list non-examples. Students can use the examples and non-examples from the probe for the third and fourth quadrants.

References

Cunningham, C. M. 2018. *Engineering in elementary STEM education: Curriculum design, instruction, learning, and assessment.* New York: Teachers College Press.

Jocz, J., and C. P. Lachapelle. 2012. *The impact of Engineering is Elementary (EiE) on students' conceptions of technology.* Boston: Museum of Science.

Keeley, P. 2016. *Science formative assessment, volume 1: 75 practical strategies for linking assessment, instruction, and learning.* 2nd ed. Thousand Oaks, CA: Corwin Press.

Lachapelle, C. P., J. D. Hertel, J. Jocz, and C. M. Cunningham. 2013. Measuring students' naïve conceptions about technology. Paper presented at the NARST Conference, Rio Grande, Puerto Rico.

National Research Council (NRC). 2012. *A framework for K–12 science education: Practices, crosscutting concepts, and core ideas.* Washington, DC: National Academies Press.

NGSS Lead States. 2013. *Next Generation Science Standards: For states, by states.* Washington, DC: National Academies Press. *www.nextgenscience.org/next-generation-science-standards.*

Image Credits

Pencil image: Cary Sneider, permission granted.

Airplane image: Cary Sneider, permission granted.

Strawberry image: CC BY-SA 3.0, Nick Youngson/Alpha Stock Images, *www.picserver.org/s/strawberry-2.html*

Horse image: CC BY-SA 3.0, Arsdelicata, *https://commons.wikimedia.org/wiki/File:Draw-Costa_Rican-2smallest.jpg*

Frog image: CC Attribution 2.0 Generic, Brian Gratwicke, *https://commons.wikimedia.org/wiki/File:Lithobates_sylvaticus_(Woodfrog).jpg*

Taco shell image: public domain, Renee Comet, U.S. Department of Health and Human Services, *https://commons.wikimedia.org/wiki/File:NCI_Visuals_Food_Taco.jpg*

Soccer ball image: CC BY-SA 3.0, Pumbaa80, *https://commons.wikimedia.org/wiki/File:Soccer_ball.svg*

Pie image: CC BY-SA 3.0, Evan-Amos, *https://commons.wikimedia.org/wiki/File:Pumpkin-Pie-Slice.jpg*

Flower image: CC BY-SA 2.0, EvaK, *https://commons.wikimedia.org/wiki/File:Nymphaea_tetragona.jpg*

Lamp image: CC BY-SA 3.0, J.Dncsn, *https://commons.wikimedia.org/wiki/File:White_lamp.JPG*

Bear image: CC BY-SA 4.0, Didier Descouens, Museum de Toulouse, *https://commons.wikimedia.org/wiki/File:Cannelle_MHNT.jpg*

Lightning image: public domain, Nico36~commonswiki, *https://commons.wikimedia.org/wiki/File:Foudre.JPG*

Eggs image: CC BY-SA 3.0, Laslovarga, *https://commons.wikimedia.org/wiki/File:American_Robin_Eggs_in_Nest.jpg*

Cell phone image: CC BY-SA 4.0, Santeri Viinamaki, *https://commons.wikimedia.org/wiki/File:Snap_on_smart_phone_20170624.jpg*

Candle image: CC BY-SA 4.0, domdomegg, *https://commons.wikimedia.org/wiki/File:Candle_flame_closeup.jpg*

Cup image: CC BY-SA 3.0, Peewack, Julius Schorzman, *https://commons.wikimedia.org/wiki/File:Cup-o-cofee-no-spoon.svg*

Volcano image: public domain, Cyrus Read, Geophysicist, U.S. Geological Survey, Alaska Volcano Observatory, *https://commons.wikimedia.org/wiki/File:Augustine_volcano_Jan_24_2006_-_Cyrus_Read.jpg*

Bird image: CC BY-SA 2.0, Tim Felce (Airwolfhound), *https://commons.wikimedia.org/wiki/File:Seagull_-_Anglesey_2009.jpg*

Balloons image: CC0 1.0, Clikr-Free-Vector-Images, *https://commons.wikimedia.org/wiki/File:Balloons-41362.png*

Hat image: CC BY-SA 4.0, Jmolina 1999, *https://commons.wikimedia.org/wiki/File:Monticristi_Straw_Hat_Optimo.jpg*

What's the Purpose of Technology?

Five friends were talking about new and improved technologies. They each had a different idea about why technologies are always being developed.

Bijou: I think technologies are developed to make life easier for everyone.

Eve: I think technologies are developed to meet people's need for clean water, food, and shelter.

Franco: I think technologies are developed to make the world a better place for people, plants, and animals.

Marisol: I think technologies are developed for all the reasons you said.

Hal: I disagree with all of you. I think technologies are developed for a different reason than what you all said.

Who do you agree with the most? ______________________ Explain your thinking.

____

____

____

____

____

____

¿Cuál Es el Propósito de Tecnología?

Cinco amigos hablaban de tecnologías nuevas y mejoradas. Cada uno tenia una idea diferente sobre por qué las tecnologías siempre se están desarrollando.

Bijou: Creo que las tecnologías están desarrolladas para hacer la vida más fácil para todos.

Eve: Creo que las tecnologías se desarrollan para que las personas tengan agua limpia, alimentos, y refugio.

Franco: Creo que las tecnologías se desarrollan para hacer el mundo un lugar mejor para las personas, las plantas, y los animales.

Marisol: Creo que las tecnologías se desarrollan por todas las razones que ustedes describieron.

Hal: Creo que las tecnologías se desarrollan por una razón diferente a la que describieron.

¿Con quién estás más de acuerdo? ______________________ Explica lo que piensas.

____

____

____

____

____

____

What's the Purpose of Technology?

Teacher Notes

Purpose
The purpose of this assessment probe is to elicit students' ideas about the purpose of technology. The probe is designed to find out if students recognize that technologies are developed to meet human needs and preserve the environment.

Type of Probe
Friendly talk

Related Key Ideas
- Technology is any modification of the natural world to fulfill human needs or desires.
- Technology includes all types of human-made systems and processes, not just modern electrical devices.
- Technology can be used to preserve and improve the environment.

Explanation
The best answer is Marisol's: "I think technologies are developed for all those reasons you said." Today, nearly everyone in developed countries has access to clean water, food, and shelter as the result of modern technologies. There are many more technologies designed to improve life in other ways, such as cars and airplanes for transportation, telephones and the internet for communication, and medicines and artificial limbs for better health. Technologies, such as renewable energy or recycling of wastes, are being developed to reduce our impact on the environment, and to preserve wild species of plants and animals.

Administering the Probe
This probe is best used with students in grades 3–12. You can extend the probe by asking students to provide several examples to support their explanation.

Connection to the Three Dimensions (NRC 2012; NGSS Lead States 2013)
- DCI: ETS2.B. Influence of Science, Engineering, and Technology on Society and the Natural World

Related Research

- Research studies have found that expanding understanding about how technology can help people greatly increases interest in STEM careers, especially for girls. For example, in the program Technovation, girls ages 10–18 identify a problem within their community that could be solved by developing a smartphone app. To evaluate the program, 653 girls who participated in the program were surveyed at least four months and up to five years after the end of the program (Vega 2016). The survey results showed that the program increased girls' interest in computer science, entrepreneurship, and business leadership. Among the Technovation participants who were now in college, 26% were majoring in computer science, a rate 65 times higher than the national average; 33% who were not in computer science were in some other STEM major, with engineering being most common—a rate twice as high as the national average.

Suggestions for Instruction and Assessment

- Use the discussion that ensues from this probe to increase students' (and especially girls') interest in technology by overcoming the stereotype that technology just includes electronics or building bridges. Help them see that technology includes a vast array of things that help people in many different ways, from medicine to preserving the environment. Be sure to use examples from their daily lives so they can relate that technology is a familiar and common element in their lives.
- You can follow this probe with an activity in which students are presented with a list of needs. For each need, they list examples of technologies that meet that need. The list of needs can include to go faster, to be safer, to be healthier, to have fun, to be warmer, to do things after dark, to be entertained, to see further, and to communicate. You can also have students come up with needs to add to the list.
- Some technologies have had unintended negative consequences such as negative impacts on the environment. Starting with a familiar technology such as automobiles, ask your students what negative environment impacts have been caused by that technology, how more careful planning might have removed or reduced those negative impacts, and what other technologies have been developed to reduce those impacts. You can do the same with other technologies such as burning coal and oil for electrical energy, and televisions, cell phones, and other electronics.
- This is a good time to introduce your students to a definition of *technology*—all of the products, processes, and systems that we develop to improve the human-made world and preserve the natural environment. Ask the students to describe how this definition applies to the technologies that they listed to meet human needs and the need for environmental protection.

References

National Research Council (NRC). 2012. *A framework for K–12 science education: Practices, crosscutting concepts, and core ideas.* Washington, DC: National Academies Press.

NGSS Lead States. 2013. *Next Generation Science Standards: For states, by states.* Washington, DC: National Academies Press. *www.nextgenscience.org/next-generation-science-standards.*

Vega, V. 2016. *Technovation lookback report: A retrospective survey of five Technovation cohorts (2010–2014).* Rockman et al. *http://iridescentlearning.org/wp-content/uploads/2014/01/TechnovationLookbackReport.pdf.*

How Do Technologies Change?

Four friends were talking about technology. They each had a different idea about how technologies change:

Fred: Technologies don't really change; they're just replaced by new technologies.

Sabrina: Technologies get more complex over time. Just look at airplanes as an example.

Dean: Technologies change in different ways. Some become more complex and others become simpler.

Adina: I think technologies change until a successful product is produced. Once the product is successful, it does not change.

Who do you agree with the most? ______________________ Explain your thinking.

____

____

____

____

____

____

¿Cómo Cambian las Tecnologías?

Cuatro amigos estaban hablando de tecnología. Cada uno tenía una idea diferente sobre cómo cambian las tecnologías:

Fred: Las tecnologías no cambian realmente; simplemente son reemplazadas por nuevas tecnologías.

Sabrina: Las tecnologías se vuelven más complejos con el tiempo. Solo mira los aviones como ejemplo.

Dean: Las tecnologías cambian de diferentes maneras. Algunos se vuelven más complejos, otros se vuelven más simples.

Adina: Creo que las tecnologías cambian hasta que se produce un producto exitoso. Una vez que el producto tiene éxito, no cambia.

¿Con quién estás más de acuerdo?________________________ Explica lo que piensas.

How Do Technologies Change?

Teacher Notes

Purpose

The purpose of this assessment probe is to elicit students' ideas about how technologies change. The probe is designed to see if students are aware of the way nearly all technologies change over time.

Type of Probe

Friendly talk

Related Key Ideas

- Technology is any modification of the natural world to fulfill human needs or desires.
- People's needs and wants change over time, as do their demands for new and improved technologies.

Explanation

The best answer is Dean's: "Technologies change in different ways. Some become more complex and others become simpler. Even the most successful technologies change as people see opportunities for improvements. The reasons for improvement vary widely and include improved performance, additional features, increased safety, and lower cost. Many technologies become more complex over time, with phones and airplanes being excellent cases in point. However, some technologies become simpler over time, such as computers, which could just be used by experts when they were first developed, and now can be used by anyone. Technologies are improved in a number of ways, including iteration (continuous improvement of a technology) and innovation (finding new uses for existing products). The idea that technologies change over time is related to the initial phase of engineering design focused on imagining new solutions, but also to the final phase of an engineering design process—to improve and optimize a solution.

Administering the Probe

This probe can be used with students in grades 3–12. You may modify the language in the answer choices for students in grades 3–5, such as changing "complex" to "complicated." You can extend the probe by asking students to give an example of a kind of technology other than phones that supports their thinking.

Connection to the Three Dimensions (NRC 2012; NGSS Lead States 2013)

- DCI: ETS2.B. Influence of Science, Engineering, and Technology on Society and the Natural World

Related Research

The Toshiba/NSTA ExploraVision competition provides an opportunity for children in grades K–12 to research technologies and envision how a current technology might be improved in the future to better meet people's needs. Since the program started in 1992, more than 450,000 children have participated. A research study of ExploraVision entries (Eisenkraft 2011) supports the claim that many children across a broad age range are able to envision how and why a given technology might change in the future. The study also found that top technology interests were medical, agricultural, energy and power, information and communication, transportation, and manufacturing. The choice of technology was found to be independent of grade level. Most students thought their innovations would have a positive societal advantage, but may have technical difficulties. Other negative side effects noted were cost and loss of jobs.

Suggestions for Instruction and Assessment

- After discussing the probe, have students identify specific examples of changes in technology.
- Have students create visual displays of technologies that have changed significantly over time.
- Ask students to identify examples of technologies that have become more complex over time. Then have them identify examples that have become simpler over time.
- Have students trace the development of the automobile. How do they think the automobile will change 20 years from now?
- Consider forming ExploraVision teams with your students. For more information about the ExploraVision competition, go to *www.exploravision.org*.

References

Eisenkraft, A. 2011. Student views of technology. *Technology and Engineering Teacher* 71 (2): 26–30.

National Research Council (NRC). 2012. *A framework for K–12 science education: Practices, crosscutting concepts, and core ideas.* Washington, DC: National Academies Press.

NGSS Lead States. 2013. *Next Generation Science Standards: For states, by states.* Washington, DC: National Academies Press. *www.nextgenscience.org/next-generation-science-standards*.

Block Diagrams

Block diagram for making corn kernels into popcorn

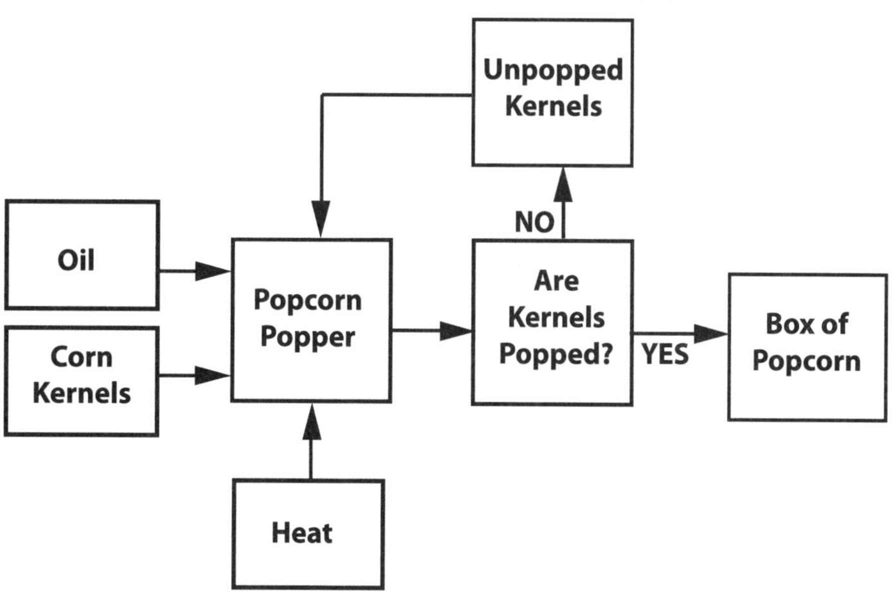

Source: Adapted from Hammack et al. 2015.

Three friends studied the block diagram shown above. They each had a different idea about what the diagram shows.

Oscar: I think the diagram shows the materials used to make a technology.

Carmela: I think the diagram shows a technology process.

Mitch: I think the diagram shows the product of technology.

Who do you agree with the most? ______________________ Explain your thinking.

____

____

____

____

____

____

Diagramas de Bloques

Diagrama de bloques para hacer granos de maíz en palomitas de maíz

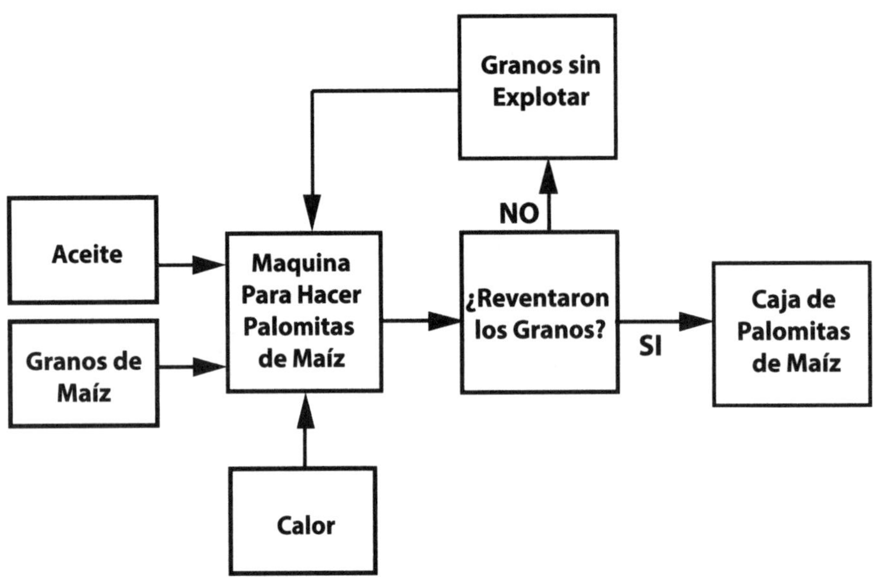

Origen: Adaptado de Hammack et al. 2015.

Tres amigos estudian el diagrama que se muestra arriba. Cada uno tenía una idea diferente sobre lo que muestra el diagrama.

Oscar: Creo que el diagrama muestra los materiales utilizados para hacer una tecnología.

Carmela: Creo que el diagrama muestra un proceso tecnológico.

Mitch: Creo que el diagrama muestra el producto de la tecnología.

¿Con quién estás más de acuerdo?____________________ Explica lo que piensas.

____

____

____

____

____

Block Diagrams

Teacher Notes

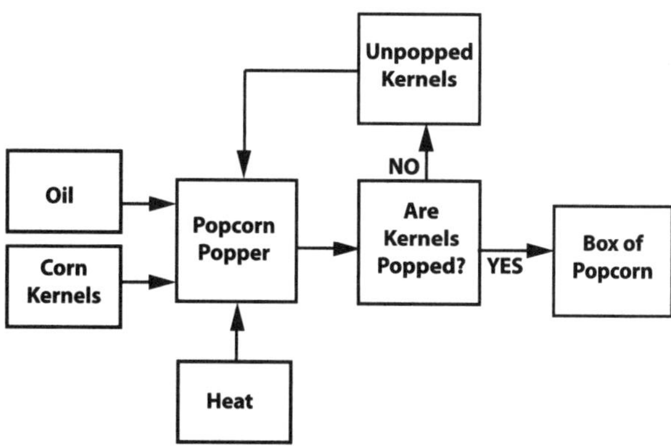

Purpose

The purpose of this assessment probe is to elicit students' ideas about block diagrams. The probe is designed to examine how students interpret a block diagram, and to determine whether they recognize that technologies include processes as well as products.

Type of Probe

Friendly talk

Related Key Ideas

- Technologies consist of products, processes, and systems.
- A block diagram is a means for showing a technological system or process.

Explanation

The best answer is Carmela's: "I think the diagram shows a technology process." The other answers are not wrong, since the diagram includes the materials used in making a technology, as well as the product. However, because the diagram as a whole illustrates the process of making popcorn, Carmela has the best answer. The block diagram shows a system in which the inputs are oil, kernels, and heat, and the output is a box of popcorn. The system also includes a means of keeping the unpopped kernels out of the box and returning them to the popper. The popcorn popping system is like many other technologies that turn natural materials into products that people need. For example, lumber is produced by cutting trees, removing them from the forest, sawing the tree into boards, then drying the boards. One way that technology processes can be described is by making block diagrams like the one shown in the probe for making popcorn out of fresh corn seeds (called kernels). Block diagrams are used often in engineering as visual models of systems. They have a common structure in which the blocks represent various components and functions, and the interconnecting lines show the relationships between blocks.

Administering the Probe

This probe is best used with students in grades 3–12. Younger students may understand the probe better if you first walk and talk through the steps in the block diagram shown in the probe. The probe can be extended by having

students explain how to use the block diagram. As reference, this type of block diagram is also known as a "flowchart," and is often used to describe processes.

Connections to the Three Dimensions (NRC 2012; NGSS Lead States 2013)

- SEP: Developing and Using Models
- SEP: Obtaining, Evaluating, and Communicating Information

Related Research

Hammack et al. (2015) reported on a four-day summer camp in which middle school students studied chemistry and engineering. One of the activities was for students to design, build, and test a separating filter for a movie theater that could separate popped and unpopped kernels so that the unpopped kernels could be fed back into the popper. The students then constructed a specific type of block diagram (called a flowchart) for their popcorn filter, similar to diagrams used by engineers in many disciplines. Results of the four-day camp were that students improved their understanding of technology and chemistry, and had more positive attitudes about engineering.

Suggestions for Instruction and Assessment

- Have each student individually describe in writing the process for popping corn, based on the block diagram in the probe. Then have pairs of students or small groups share their descriptions with each other. Have them discuss how their descriptions are alike or different.
- As students learn about engineering and technology, they should have opportunities to interpret and create block diagrams. Have students generate a list of everyday products or processes they are familiar with. Ask them to create block diagrams to describe the inner workings of the product or process, and its inputs and outputs to its environment.
- Although the probe focuses on a block diagram for making something (popcorn), you can discuss with your students that technologies include other kinds of processes, such as chemical processes for producing medicines, production of steel from iron ore, or making bread in a commercial bakery.
- For students in the younger grades, a way to introduce block diagrams is to relate it to the teacher's use of board drawings to illustrate different concepts. A board drawing, which should be familiar to students, is a visual way of relating different concepts. Similarly, a block diagram is a visual way to show how different parts of a technological system relate to each other.

References

Hammack, R., T. A. Ivey, J. Utley, and K. A. High. 2015. Effect of an engineering camp on students' perceptions of engineering and technology. *Journal of Pre-College Engineering Education Research* 5 (2): 10–21.

National Research Council (NRC). 2012. *A framework for K–12 science education: Practices, crosscutting concepts, and core ideas.* Washington, DC: National Academies Press.

NGSS Lead States. 2013. *Next Generation Science Standards: For states, by states.* Washington, DC: National Academies Press. *www.nextgenscience.org/next-generation-science-standards.*

Technology, System, or Both?

Two cooks were cutting vegetables with a chopping knife. They each had a different idea about the knife:

Susanna: I think the chopping knife is a technological system.

Armand: I think the chopping knife is a technology, but not a system.

Who do you agree with the most? ______________________ Explain why you agree.

____

____

____

____

____

____

¿Tecnología, Sistema o Ambos?

Dos cocineros usaban cuchillos para picar vegetales. Cada uno tenía una idea diferente sobre el cuchillo:

Susanna: Creo que el cuchillo para picar es un sistema tecnológico.

Armand: Creo que el cuchillo para picar es una tecnología, pero no un sistema.

¿Con quién estás más de acuerdo? ______________________ Explica lo que piensas.

____

____

____

____

____

____

Technology, System, or Both?

Teacher Notes

Purpose

The purpose of this assessment probe is to elicit students' ideas about technological systems. The probe is designed to determine if students are able to envision how a system is made up of parts and can do something that the parts alone cannot do.

Type of Probe

Opposing views

Related Key Ideas

- Technologies consist of products, processes, and systems.
- A system is composed of two or more parts, and its function differs from that of the individual parts.

Explanation

The best answer is Susanna's: "I think the chopping knife is a technological system." While the knife is an example of a technology; many technologies are also technological systems. A *technological system* is a group of parts that are related to each other so that the system as a whole behaves in ways that the individual parts do not. Technological systems can be products, processes, or both. A knife is a simple technological system that consists of three parts: a cutting blade, handle, and metal fasteners (called rivets) that hold the other parts together. Each part alone cannot do what the knife does: The handle and the rivets cannot chop, and the blade by itself cannot be held safely or balanced for chopping.

Administering the Probe

This probe is best used with students in grades 3–12. You can show students a picture of a kitchen knife that shows the blade, handle, and rivets as well as a video of a cook using a chopping knife. Make sure students know what a system is before using this probe.

Connection to the Three Dimensions (NRC 2012; NGSS Lead States 2013)

- CCC: Systems and System Models

Related Research

- Silk and Schunn (2008) Identified engineering concepts of systems and optimization as foundational to the discipline of engineering education. They then reviewed existing literature and concluded that "a focus on the core engineering concepts of systems and optimization in K–12 engineering education has the potential to provide a solid foundation for continued study of engineering at the undergraduate level and beyond" (p. 1).
- Mehalik, Doppelt, and Schunn (2008) presented a unit in which students designed and built electrical alarm systems to 26 classes (585 students) of eighth graders, and compared the results with 20 classes (466 students) of eighth graders who learned the same material through a scripted approach. In the scripted approach, the students do the activity with the same materials, but their work is heavily scaffolded to help them discover scientific concepts rather than encourage them to design the system. The students were assessed on science content that was included in both teaching methods. The students who learned by designing systems had a higher achievement in core science concepts, engagement, and retention than the students who learned through the scripted approach. The systems-design approach was most helpful to low-achieving African American students.
- Research conducted with an elementary science program (Science Curriculum Improvement Study) showed that elementary students may believe that a system of objects must be doing something to be considered a system. They also believed that a system that loses a part is still the same system (Hill and Redden 1985).

Suggestions for Instruction and Assessment

- Prior to administering this probe, use the "Is It a System?" probe in *Uncovering Student Ideas in Science, Volume 4:* to find out how students think about the concept of systems (Keeley and Tugel 2009).
- Have students choose a simple technological system (other than a knife) made from two or more parts. Have them sketch the parts to show how they relate to each other. Ask students what the assembled object can do that the parts alone cannot do.
- Go on a "parts and wholes" technological systems walk, in which students identify different technological systems. They describe the parts that make up the system and explain how each part contributes to the system as a whole.
- Compare technological systems with natural systems. Have students examine a variety of familiar technological systems (e.g., bicycle, pencil sharpener, flashlight). Describe what the system does and how the components interact and contribute to the system as a whole. Then do the same with a natural system, such as an ecosystem, human body, weather system, or the solar system.

References

Hill, D. M., and M. G. Redden. 1985. An investigation of the system concept. *School Science and Mathematics* 85 (3): 233–239.

Keeley, P., and J. Tugel. 2009. *Uncovering student ideas in science, volume 4: 25 new formative assessment probes.* Arlington, VA: NSTA Press.

Mehalik, M. M., Y. Doppelt, and C. D. Schunn. 2008. Middle-school science through design-based learning versus scripted inquiry: Better overall science concept learning and equity gap reduction. *Journal of Engineering Education* 97 (1): 71–85.

National Research Council (NRC). 2012. *A framework for K–12 science education: Practices, crosscutting concepts, and core ideas.* Washington, DC: National Academies Press.

NGSS Lead States. 2013. *Next Generation Science Standards: For states, by states.* Washington, DC: National Academies Press. *www.nextgenscience.org/next-generation-science-standards.*

Silk, E. M., and C. D. Schunn. 2008. Core concepts in engineering as a basis for understanding and improving K–12 engineering education in the United States. Report to the National Academy of Engineering Committee on Understanding and Improving K–12 Engineering Education in the United States. *http://elisilk.net/research/SilkSchunn2008a-NAE-FinalDraft.pdf.*

Systems Within Systems

Can a technological system be a part of another system? Choose the statement that best matches your thinking about systems.

___ **A.** A technological system is always part of a larger system.

___ **B.** A technological system is sometimes part of a larger system.

___ **C.** A technological system can never be part of a larger system.

Which statement do you agree with the most? ________________ Explain your thinking.

____

____

____

____

____

____

Sistemas Dentro de Sistemas

¿Puede un sistema tecnológico ser parte de otro sistema? Elija la declaración que mejor se adapte a su pensamiento sobre los sistemas.

___ **A.** Un sistema tecnológico siempre es parte de un sistema más grande.

___ **B.** Un sistema tecnológico es a veces parte de un sistema más grande.

___ **C.** Un sistema tecnológico nunca puede ser parte de un sistema más grande.

¿Con qué declaración está más de acuerdo?__________________ Explica lo que piensas.

____

____

____

____

____

____

Systems Within Systems

Teacher Notes

Purpose

The purpose of this assessment probe is to elicit students' ideas about technological systems. The probe is designed to determine the extent to which students recognize that all systems are nested within and interconnected with larger systems.

Type of Probe

Always, sometimes, never

Related Key Ideas

- Technologies consist of products, processes, and systems.
- A system is composed of two or more parts, and its function differs from that of the individual parts.
- Technological systems are nested within larger systems.

Explanation

The best answer is A: A technological system is always part of a larger system. For example, an athletic shoe worn by a basketball player has several different parts, each designed to give the wearer comfort and support. The shoe is just one part of a basketball player's clothing, which in turn is part of a larger system that includes the ball and net, the court, and chairs or benches for the spectators. If it is a high school game, the basketball system includes uniforms and pom-poms used by cheerleaders, school buses that transport teams to different cities, and newspapers that report the results.

In practice, it is sometimes a useful simplification to define and think of a system as being by itself so that it can be analyzed or operated fairly independently of outside influences. For example, in a controlled scientific experiment, scientists often go to great lengths to isolate a system under study from external influences that might disturb it during the experiment. This is an example of a situation in which it is useful to think of a system by itself. In general, though, it is important in engineering to encourage students to consider how a system fits in with other systems, since a change in one system may cause changes in related systems.

Administering the Probe

This probe is best used with students in grades 6–12, after they learn that a system is made up of interacting parts. You can extend the probe by having students provide examples to support their answer choice.

Connection to the Three Dimensions (NRC 2012; NGSS Lead States 2013)

- CCC: Systems and System Models

Related Research

- Dalrymple, Sears, and Evangelou (2010) found significant pre-post assessment differences in first-year college students' understanding of system interconnectivity and their abilities to redesign solutions after doing a scaffolded "dissection" activity in which they took a product apart to examine its subsystems.
- Silk and Schunn (2008) Identified engineering concepts of systems and optimization as foundational to the discipline of engineering education. They then reviewed existing literature and concluded that "a focus on the core engineering concepts of systems and optimization in K–12 engineering education has the potential to provide a solid foundation for continued study of engineering at the undergraduate level and beyond" (p. 1).
- Mehalik, Doppelt, and Schunn (2008) presented a unit in which students designed and built electrical alarm systems to 26 classes (585 students) of eighth graders, and compared the results with 20 classes (466 students) of eighth graders who learned the same material through a scripted approach. In the scripted approach, the students do the activity with the same materials, but their work is heavily scaffolded to help them discover scientific concepts rather than encourage them to design the system. The students were assessed on science content that was included in both teaching methods. The students engaged in designing systems had a higher achievement in core science concepts, engagement, and retention than the students who learned through the scripted approach. The systems design approach was most helpful to low-achieving African American students.

Suggestions for Instruction and Assessment

- Starting with a brick, have students fill in the blanks in the following sentence (possible answers are in parentheses): A brick is part of a __________ (wall), which is part of a __________ (building), which is part of a __________ (city). Then challenge them to select a simple technological object and list how many systems within systems they can come up with. For example: A handle is part of a faucet, which is part of a sink, which is part of the kitchen, which is part of a house, which is part of a neighborhood, which is part of a city.
- Ask the students to name the major systems in a car, such as the body, engine, and tires. Then choose one of the systems and ask them to name some of its subsystems, then the parts that make up those subsystems, and so on.
- Compare technological systems with natural systems to see how both are made up of systems within systems. For example, a cell is part of a system of tissues, which are part of an organ, which is part of an organ system, which is part of an organism, which is part of a population of organisms, which is part of an ecosystem.
- Starting with the basketball player's shoe, describe the shoe as a system with interacting parts. Then describe the shoe as one part of a basketball player's clothing, and

so on. See how far the students can get at nesting systems within systems.

References

Dalrymple, O., D. Sears, and D. Evangelou. 2010. Evaluating the motivational and learning potential of an instructional practice for use with first year engineering students. *Proceedings of the Annual Meeting of the American Society for Engineering Education*, Louisville.

Mehalik, M. M., Y. Doppelt, and C. D. Schunn. 2008. Middle-school science through design-based learning versus scripted inquiry: Better overall science concept learning and equity gap reduction. *Journal of Engineering Education* 97 (1): 71-85.

National Research Council (NRC). 2012. *A framework for K–12 science education: Practices, crosscutting concepts, and core ideas.* Washington, DC: National Academies Press.

NGSS Lead States. 2013. *Next Generation Science Standards: For states, by states.* Washington, DC: National Academies Press. *www.nextgenscience.org/next-generation-science-standards.*

Silk, E. M., and C. D. Schunn. 2008. Core concepts in engineering as a basis for understanding and improving K–12 engineering education in the United States. Report to the National Academy of Engineering Committee on Understanding and Improving K–12 Engineering Education in the United States. *http://elisilk.net/research/SilkSchunn2008a-NAE-FinalDraft.pdf.*

Section 2
What Is Engineering?

Key Ideas Matrix for Probes #8–#16

PROBES	#8 Who Engineers?	#9 Who Can Become an Engineer?	#10 Team Players?	#11 Working Together to Save Lives	#12 How Are Science and Engineering Similar?	#13 Is Engineering Creative?	#14 Reasons for Success	#15 Is Engineering Experimental?	#16 Is It Rocket Science?
GRADE-LEVEL USE →	**3–12**	**3–12**	**3–12**	**3–12**	**5–12**	**3–12**	**3–12**	**3–12**	**6–12**
RELATED KEY IDEAS ↓									
Engineers include people from many different ethnic and cultural backgrounds.	X				X				
Engineering as a profession is open to people with a wide variety of interests and capabilities.	X	X			X				
Engineers design products, processes, and systems that meet people's needs.		X							
Engineering requires scientific and mathematical thinking.		X						X	
Engineers create solutions for problems that deeply affect people, including some that save lives.				X					
Engineers nearly always work in teams, often including people from other professions.			X	X					
Science and engineering complement each other and help each other progress.									X
Engineering and science share many common practices.					X				
Engineering is a highly creative process.						X			
Both engineering and science require that arguments be based on evidence and logical reasoning.							X		
Both scientists and engineers perform experiments that involve control and manipulation of variables.							X	X	
Scientists discover new knowledge, and engineers create solutions to meet people's needs.									X

Teaching and Learning Considerations

According to a report from the National Science Board (2015), STEM skills are needed in a wide variety of jobs. However, misconceptions about the nature of engineering abound, leading many students away from engineering before they have even had a chance to learn what it is about, and how powerful it can be for making a better world. The purpose of the probes in this section is to provide an opportunity to learn what your students think about the practice of engineering—what do engineers actually do? These probes will spark conversations to better understand and appreciate the field.

This section begins with a series of probes about students' initial ideas concerning who can become an engineer, and what motivates engineers. Other probes reveal students' ideas about the relationship between scientists and engineers, how engineers work together in teams, and the nature of an engineering design process as highly creative, logical, and scientific.

A detailed summary of what students in grades K–2, 3–5, 6–8, and 9–12 can be expected to know and be able to do with respect to engineering design and skills, such as troubleshooting, invention, and innovation, can be found in the *Standards for Technological Literacy* (ITEEA 2007) on pages 89–138.

A considerable amount of research underlies the probes in this section, and we urge readers to look at the summaries in the probes' Related Research sections for clues about what will help their students recognize the value of engineering and the wide diversity of people in the engineering profession. For many students, the practical applications of engineering provide a motivation for engaging in the STEM fields, and build problem-solving and designs skills that have lifelong value no matter what vocation they pursue.

References

International Technology and Engineering Educators Association (ITEEA). 2007. *Standards for Technological Literacy: Content for the study of technology.* 3rd ed. Reston, VA: ITEEA. *www.iteea.org/File.aspx?id=67767.*

National Science Board. 2015. *Revisiting the STEM workforce: A companion to Science and Engineering Indicators 2014.* Arlington, VA: National Science Foundation. *www.nsf.gov/pubs/2015/nsb201510/nsb201510.pdf.*

Who Engineers?

Draw a picture of an engineer(s) doing work. Include some of the things engineers use.

Explain your drawing. What kinds of people are engineers and what do they do?

¿Quiénes Son los Ingenieros?

Haga un dibujo de un ingeniero(s) haciendo su(s) trabajo. Incluya algunas de las cosas que usan los ingenieros.

Explica tu dibujo. ¿Qué tipo de personas son ingenieros y qué hacen?

Who Engineers?

Teacher Notes

Draw a picture of an engineer(s) doing work. Include some of the things engineers use.

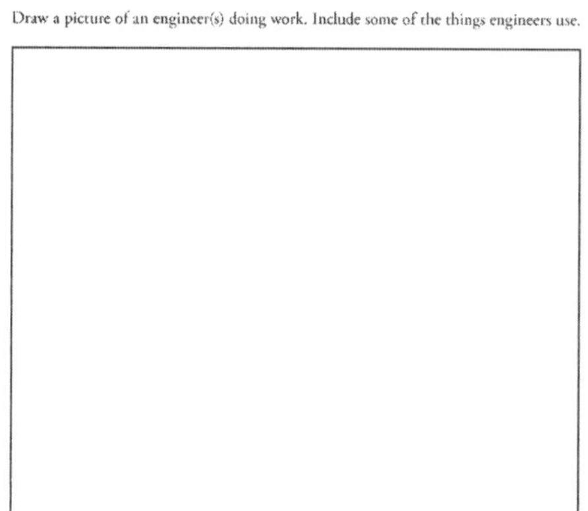

Purpose

The purpose of this assessment probe is to elicit students' ideas about engineers. The probe is designed to reveal how students think about who is an engineer and the kind of work engineers do.

Type of Probe

Draw a picture

Related Key Ideas

- Engineers include people from many different ethnic and cultural backgrounds.
- Engineering as a profession is open to people with a wide variety of interests and capabilities.

Explanation

There is no single best answer to this probe. The Draw-an-Engineer task is commonly used to assess what students know about engineering careers, and whether or not they have any misconceptions, such as the commonly held idea that only white men are engineers. Students' pictures and descriptions should ideally show men and women of different races, ethnicities, and cultures involved in a variety of types of engineering.

In addition to noticing gender, race, and other characteristics, examine your students' drawings for clues about what kinds of activities they believe engineers engage in. What objects are they holding? What are they wearing? What appears to distinguish them as engineers? Important features to look for, and discuss afterward with students, are engineers interacting with other people such as colleagues in teams, users of the products they are designing in their own surroundings, or any other situation. The key point is to highlight that engineers don't work solo hiding in a cubicle or lab!

Administering the Probe

This probe is best used with students in grades 3–12. Note that this probe differs from others in the *Uncovering Student Ideas* series as it does not include a response that the students can select. Remind students that their drawings should show two things: (1) the type of person who is an engineer, and (2) what the engineer is doing.

Connection to the Three Dimensions (NRC 2012; NGSS Lead States 2013)

- Connection to the Nature of Science: Science Is a Human Endeavor

Related Research

- Several researchers have used the Draw-an-Engineer test. Thompson and Lyons (2006) paired graduate engineering students with elementary and middle school teachers to help them develop lessons on engineering for fourth and fifth graders. After three years, all sixth-grade students in the school were given the Draw-an-Engineer test, including 44 students who had participated in the program and 122 controls who had not. Ten students from each group then participated in individual interviews. The Draw-an-Engineer assessment showed that students who participated in the program developed clearer perceptions of engineering than had the controls. These findings were supported by quotes from the interview data, which showed that the graduate students were influential in helping the students develop more realistic images of engineers.
- Oware, Capobianco, and Diefes-Dux (2007) studied 18 gifted students who had completed third or fourth grade the prior year, and who were enrolled in summer engineering classes. Analysis of their Draw-an-Engineer tests showed that half of the students knew an engineer and 30% had taken engineering classes previously. The test showed that some students had common misconceptions of engineers, but others understood that engineers design or plan things. Students drew engineers using safety items, computers, models, blueprints, and plans. Students also described things engineers considered, such as whether their product worked and motives for their work.
- Rivale et al. (2011) administered the Draw-an-Engineer test to 335 fifth-grade students before and after engaging in engineering activities taught by engineering graduate students and learning about key messages concerning engineering that the graduate students derived from a study by the National Academy of Engineering (*Changing the Conversation,* National Academy of Engineering 2008). Analysis of the results showed that the messages made a significant difference in students' knowledge of what engineers do and that girls were more likely to draw a female engineer if they were taught by a female graduate teaching fellow.
- Weber et al. (2011) developed a coding system for elementary students' drawings of engineers. Although intended for researchers to use consistent methods in analyzing the results of studies, teachers may find their coding system useful.

Suggestions for Instruction and Assessment

- After students complete this probe, examine their drawings and descriptions to gain insight into what they think about who engineers, and what engineers do. To gain further insights, spark a discussion by asking your students what they were trying to show with their drawings.
- High-quality curricula should make connections to geography and culture, and a variety of engineering fields connected to physical, life, and Earth and space sciences.
- After this activity, show your students several short videos to illustrate the wide variety of engineering and related professions (you should be able to find videos by searching on the internet).
- For your own interest, you may want to check out "PK–12 Engineering in the Classroom," a blog that includes links to several

excellent videos demonstrating engineering in the classroom. Find it along with many other resources on the LinkEngineering website at *www.linkengineering.org/Explore/LE_Blog/58354.aspx*.

References

National Academy of Engineering. 2008. *Changing the conversation: Messages for improving public understanding of engineering.* Washington, DC: National Academies Press. *https://doi.org/10.17226/12187.*

National Research Council (NRC). 2012. *A framework for K–12 science education: Practices, crosscutting concepts, and core ideas.* Washington, DC: National Academies Press.

NGSS Lead States. 2013. *Next Generation Science Standards: For states, by states.* Washington, DC: National Academies Press. *www.nextgenscience.org/next-generation-science-standards.*

Oware, E., B. Capobianco, and H. Diefes-Dux. 2007. Gifted students' perceptions of engineers? A study of students in a summer outreach program. Paper presented at 2007 ASEE Annual Conference & Exposition, Honolulu. *https://peer.asee.org/2656.*

Rivale, S., J. Yowell, J. Aiken, S. Adhikary, D. Knight, and J. F. Sullivan. 2011. Elementary students' perceptions of engineers. Paper presented at 2011 ASEE Annual Conference & Exposition, Vancouver, BC. *https://peer.asee.org/17833.*

Thompson, S., and J. Lyons. 2006. Investigating the long-term impact of an engineering-based GK–12 program on students' perceptions of engineering. Paper presented at 2006 ASEE Annual Conference & Exposition, Chicago. *https://peer.asee.org/1142.*

Weber, N., D. Duncan, M. Dyehouse, J. Strobel, and H. A. Diefes-Dux. 2011. The development of a systematic coding system for elementary students' drawings of engineers. *Journal of Pre-College Engineering Education Research* 1 (1): 49–62.

Who Can Become an Engineer?

Tanya: My neighbor told me that anyone can become an engineer.

Anna: That's definitely true. My sister is an engineer, but when she was young she was only interested in history and community service. In high school, she learned that engineers solve real problems in society, so she became a civil engineer to work on affordable housing.

Marisol: That's exciting! I can become an engineer and solve real problems! I heard that engineers need science and math so now I see why they're such valuable classes.

Leon: Anyone can become an engineer, even if they don't study science and math. You can learn science and math after you become an engineer.

Who do you agree with more—Marisol or Leon? ____________________ Explain why you agree.

____

____

____

____

____

¿Quién Puede Ser Ingeniero?

Tanya: Alguien me dijo que cualquiera puede ser ingeniero.

Anna: Eso es definitivamente cierto. Mi hermana es una ingeniera, pero cuando era joven solo le interesaba la historia y el servicio comunitario. En la escuela secundaria aprendió que los ingenieros resuelven problemas reales en la sociedad, por lo tanto se convirtió en ingeniera civil para trabajar en viviendas asequibles.

Marisol: ¡Es emocionante! ¡Puedo convertirme en ingeniero y resolver problemas reales! Pero, escuché que tienes que estudiar ciencias y matemáticas como parte de la capacitación en ingeniería.

Leon: Cualquiera puede convertirse en ingeniero, incluso si no estudian ciencias y matemáticas. Puedes aprender ciencias y matemáticas después de convertirte en ingeniero.

¿Con quién estás más de acuerdo: Marisol o Leon?__________________ Explica por qué estás de acuerdo.

____

____

____

____

____

Who Can Become an Engineer?

Teacher Notes

Purpose

The purpose of this assessment probe is to elicit students' ideas about who can become an engineer. The probe is designed to reveal whether students recognize that anyone can become an engineer as long as they study science and math, have a willingness to learn new ideas, and want to help people.

Type of Probe

Follow the dialogue

Related Key Ideas

- Engineering as a profession is open to people with a wide variety of interests and capabilities.
- Engineers design products, processes, and systems that meet people's needs.
- Engineering requires scientific and mathematical thinking.

Explanation

The best answer is Marisol's: "That's exciting! I can become an engineer and solve real problems! I heard that engineers need science and math so now I see why they're such valuable classes." It is a common misconception that college admissions officers favor male applicants to engineering programs. Anna's story about her sister highlights that engineering is equally suited to all genders, and diversity brings valuably different awareness to understanding user needs and solutions. Although students do need to study science and math and do reasonably well, engineering programs are not just for the highest-achieving students. It is also important for students to have a desire both to learn and to help people, society, or the environment.

Administering the Probe

This probe is best used with students in grades 3–12. For younger students, you may wish to first explain briefly that a *civil engineer* designs physical structures such as buildings or bridges. For older students, the probe can be extended by asking students to list what they consider to be the most important characteristics and abilities for students to be selected for an engineering program at a two-year or four-year college.

Connection to the Three Dimensions (NRC 2012; NGSS Lead States 2013)

- Connection to the Nature of Science: Science Is a Human Endeavor

Related Research

- According to a report from the National Science Board (2015), STEM skills are needed in a wide variety of jobs. "The report's take-home message is that STEM knowledge and skills enable both individual opportunity and national competitiveness. ... Ensuring access to high quality education and training experiences for all students at all levels and for all workers at all career stages is absolutely essential" (Soergel 2015).
- In 2014, and again in 2018, a nationally representative sample of eighth graders took the National Assessment of Educational Progress in Technology and Engineering Literacy (NAEP-TEL), in which they were asked to solve real-world scenarios involving technology and engineering challenges. In both administrations of the assessment (to 20,500 eighth graders in 2014 and 15,400 eighth graders in 2018), girls scored significantly higher than boys (The Nation's Report Card 2018).
- A report by the Extraordinary Women Engineers Coalition (2005) describes a research project to learn about high school girls' perceptions of engineering. The researchers conducted focus groups with 85 girls in grades 9–12, follow-up online surveys with 74 of the focus group participants, and online surveys with 165 additional girls. Several teachers, college students, and engineers were also interviewed as part of the study. Key findings were that as a group, high school girls are interested in careers that are enjoyable, have a good working environment, make a difference in people's lives, offer a good salary, and are flexible. Although these characteristics describe careers in engineering, few girls were aware of it. Instead, most high school girls believed that engineering is just for people who love math and science.

Suggestions for Instruction and Assessment

- Have students interview an engineer about his or her middle and high school background, including what interested or prepared them to go into engineering.
- Assign students to research the accomplishments of famous women engineers, such as Grace Hopper (1906–1992), a computer scientist and U.S. Navy rear admiral, whose work led to the development of high-level computer languages still used today; Lynn Conway (1938–), a highly recognized woman and one of the first transgender engineers, who pioneered chip design methodology that powers ALL of our modern computing and electronics; or Hedy Lamarr (1914–2000), an actress and inventor whose ideas are incorporated in modern cellphone technology. Students can explore the following links, or search for "Famous Women Engineers" to find many more.
    - *https://en.wikipedia.org/wiki/Grace_Hopper*
    - *https://en.wikipedia.org/wiki/Lynn_Conway*
    - *https://en.wikipedia.org/wiki/Hedy_Lamarr*
- Have students identify two-year and four-year colleges in their state that offer programs in engineering or engineering technology, and then have them search their websites for information, such as requirements for admission and names of required courses.
- For students who may be interested in pursuing an engineering career, ask them

to list things they can do or improve on now to prepare them for an engineering program.

References

Extraordinary Women Engineers Coalition. 2005. *Extraordinary women engineers final report.* Produced by a team from the American Association of Engineering Societies, the American Society of Civil Engineers, and the WGBH Educational Foundation. *http://kellrobotics.org/files/pdf/EWE.pdf.*

National Research Council (NRC). 2012. *A framework for K–12 science education: Practices, crosscutting concepts, and core ideas.* Washington, DC: National Academies Press.

National Science Board. 2015. *Revisiting the STEM workforce: A companion to Science and Engineering Indicators 2014.* Arlington, VA: National Science Foundation *www.nsf.gov/pubs/2015/nsb201510/nsb201510.pdf.*

NGSS Lead States. 2013. *Next Generation Science Standards: For states, by states.* Washington, DC: National Academies Press. *www.nextgenscience.org/next-generation-science-standards.*

Soergel, A. 2015. Report: Non-STEM fields increasingly require STEM skills. *U.S. News and World Report,* April 21. *www.usnews.com/news/stem-solutions/articles/2015/04/21/national-science-board-report-suggests-non-stem-fields-now-require-stem-skills.*

The Nation's Report Card. 2018. Technology and engineering literacy: An innovative assessment in an era of rapid technological change. *www.nationsreportcard.gov/tel_2018_highlights.*

Team Players?

Four friends on a soccer team were trying out a new style of soccer ball. As they tried the new ball, they wondered if the engineers that designed it also worked in teams. They each had different ideas about how engineers did their work:

Sarah: When I get to college I don't plan to study engineering because I prefer to work with a team of people. I think most engineers work alone.

Franz: What makes you think engineers don't work with other people? I think engineers work with other engineers in the same field.

Marvin: I think engineers work on teams with other engineers in the same field, and engineers in other fields.

Seneca: I think engineers form teams with people in all sorts of different fields—anyone who has knowledge or experience about the problem they're working on.

Who do you think has the best description of how engineers work? ________________
Explain your thinking.

____

____

____

____

____

¿Jugadores de Equipo?

Cuatro amigos en un equipo de fútbol estaban probando un nuevo estilo de balón de fútbol. Mientras probaban la nueva pelota, se preguntaban si los ingenieros que la diseñaron también trabajaban en equipos. Cada uno tenía ideas diferentes sobre cómo los ingenieros hicieron su trabajo:

Sarah: Cuando llegue a la universidad no planeo estudiar ingeniería porque prefiero trabajar con un equipo de personas. Creo que la mayoría de los ingenieros trabajan solos.

Franz: ¿Qué te hace pensar que los ingenieros no trabajan con otras personas? Creo que los ingenieros trabajan con otros ingenieros en el mismo campo.

Marvin: Yo creo que los ingenieros trabajan en equipos con otros ingenieros en el mismo campo, e ingenieros en otros campos.

Seneca: Yo creo que los ingenieros forman equipos con personas en todo tipo de campos diferentes, cualquiera que tenga conocimiento o experiencia sobre el problema en el que están trabajando.

¿Quién crees que tiene la mejor descripción de cómo trabajan los ingenieros? ____________________ Explica lo que piensas.

____

____

____

____

____

Team Players?

Teacher Notes

Purpose

The purpose of this assessment probe is to elicit student' ideas about engineers and teamwork. The probe is designed to see if students recognize that different types of engineers and other professionals must work as members of a team to accomplish a task.

Type of Probe

Friendly talk

Related Key Idea

- Engineers nearly always work in teams, often including people from other professions.

Explanation

The best answer is Seneca's: "I think Engineers form teams with people in all sorts of different fields—anyone who has knowledge or experience about the problem they're working on." A team can include engineers in the same field, engineers in other fields, non-engineers whose expertise relates to the problem that needs to be solved, and people who may use the products or systems that the team is developing. For example, consider a team designing a new and improved running shoe. Mechanical engineers design the shoe's structure so it provides both firm support and cushioning. Materials engineers select the best materials to use and how to fasten them together. Podiatrists (doctors who specialize in feet) suggest improvements to avoid injuries. The team should also include runners who can describe what they are looking for when they purchase a new pair of shoes, and the runners need to be able to test prototypes to see how they perform in a race. And since people are users of technology, teams often have experts with backgrounds related to human behavior such as psychology, anthropology, and sociology.

Administering the Probe

This probe is best used with students in grades 3–12. You can extend the probe by having students describe an example to support their answer choice.

Connections to the Three Dimensions (NRC 2012; NGSS Lead States 2013)

- DCI: ETS2.A: Interdependence of Science, Engineering, and Technology
- Connection to the Nature of Science: Science Is a Human Endeavor

Related Research

- A report by the Extraordinary Women Engineers Coalition (2005) describes a research project to learn about high school girls' perceptions of engineering. The researchers conducted focus groups with 85 girls in grades 9–12, follow-up online surveys with 74 of the focus group participants, and online surveys with 165 additional girls. Several teachers, college students, and engineers were also interviewed as part of the study. Key findings were that high school girls do not understand what engineering is and believe it is just for people who love math and science. As a group, high school girls are interested in careers that are enjoyable, have a good working environment, make a difference in people's lives, offer a good salary, and are flexible. Although these characteristics describe careers in engineering, few girls were aware of it. Instead, most high school girls believed that engineering is just for people who love math and science.
- Pallis and McNitt-Gray (2013) evaluated a weeklong summer residential sports science and engineering academy for girls entering grades 9 through 11. Get SSET (Sport Science, Engineering and Technology) was first developed in 2003, and has been presented a number of times since then by female university faculty and sports science professionals. Based on the student evaluations, young women with a strong interest in both sports and engineering obtain the most benefit. The young women reported having a connection with the other women in the program, since they feel isolated in school by their unique interest in both sports and STEM.

Suggestions for Instruction and Assessment

- Ask students to think of teams that they have been part of. What was the team attempting to accomplish? Who were the other members of the team, and how did the different capabilities of the team members contribute to the team's overall accomplishments?
- Have students choose their favorite sport. Ask them to describe what they think would be an ideal team to help improve the equipment needed to play that sport.
- Have a discussion about the role of non-engineers in developing a product or process, specifically the people who will benefit from the design. What would be the effect on the product or process if the people who benefit from the design are not involved in the development?
- If your students are especially interested in athletic shoe design, recommend that they research the history of athletic shoe development, beginning in the 1970s, noting the variety of expertise that has been involved.

References

Extraordinary Women Engineers Coalition. 2005. *Extraordinary women engineers final report.* Produced by a team from the American Association of Engineering Societies, the American Society of Civil Engineers, and the WGBH Educational Foundation. *http://kellrobotics.org/files/pdf/EWE.pdf.*

National Research Council (NRC). 2012. *A framework for K–12 science education: Practices,*

crosscutting concepts, and core ideas. Washington, DC: National Academies Press.

NGSS Lead States. 2013. *Next Generation Science Standards: For states, by states.* Washington, DC: National Academies Press. *www.nextgenscience.org/next-generation-science-standards.*

Pallis, J. M., and J. L. McNitt-Gray. 2013. Using sports to attract young women into engineering. Paper presented at 2013 ASEE Annual Conference & Exposition, Atlanta. *https://peer.asee.org/22724.*

Working Together to Save Lives

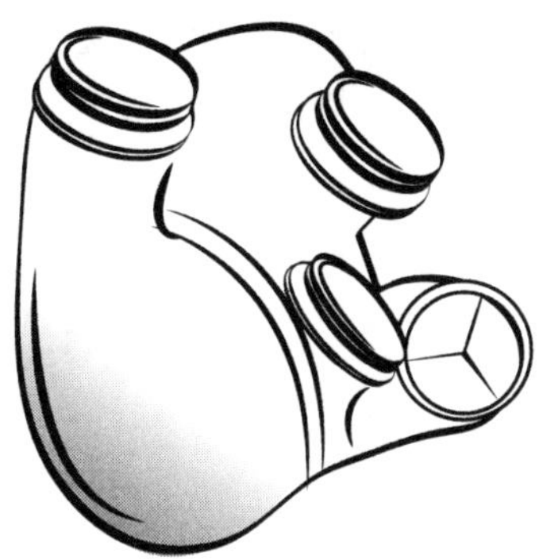

An artificial heart is an example of technology saving lives. It shows that engineers and scientists often work in teams of people with many different skills. Put an X next to any professions that might be valuable for teams developing artificial heart technology.

___ **A.** Cardiologist

___ **B.** Materials scientist

___ **C.** Surgeon

___ **D.** Manufacturing engineer

___ **E.** Biochemist

___ **F.** Mechanical engineer

___ **G.** Veterinarian

___ **H.** Intensive care nurse

___ **I.** Psychologist

___ **J.** Electrical engineer

___ **K.** Biomedical engineer

___ **L.** Physical therapist

___ **M.** Mathematician

Explain your thinking. Why do engineers and scientists work with others?

____

____

____

____

Trabajando Juntos Para Salvar Vidas

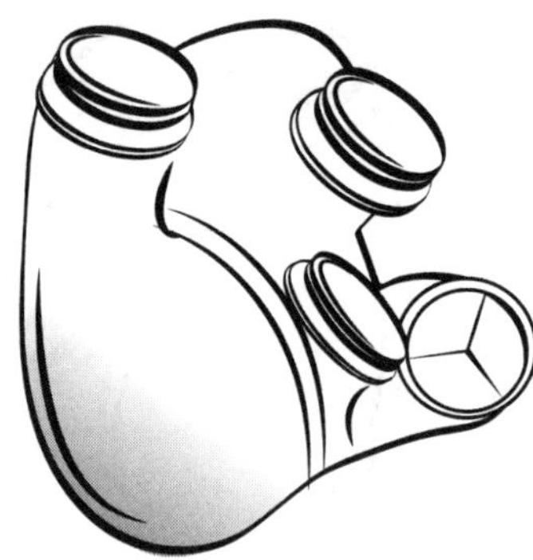

Un corazón artificial es un ejemplo de tecnología que salva vidas. Demuestra que los ingenieros y científicos a menudo trabajan en equipos de personas con muchas habilidades diferentes. Ponga una X al lado de cualquier profesión que pueda ser valiosa para los equipos que desarrollan tecnología de corazón artificial.

___ **A.** Cardiólogo

___ **B.** Científico de materiales

___ **C.** Cirujano

___ **D.** Ingeniero de fabricación

___ **E.** Bioquímico

___ **F.** Ingeniero mecánico

___ **G.** Veterinario

___ **H.** Enfermera de cuidados intensivos

___ **I.** Psicólogo

___ **J.** Ingeniero eléctrico

___ **K.** Ingeniero biomédico

___ **L.** Fisioterapeuta

___ **M.** Matemático

Explica lo que piensas. ¿Por qué los ingenieros y científicos trabajan con otros?

____

____

____

____

Working Together to Save Lives

Teacher Notes

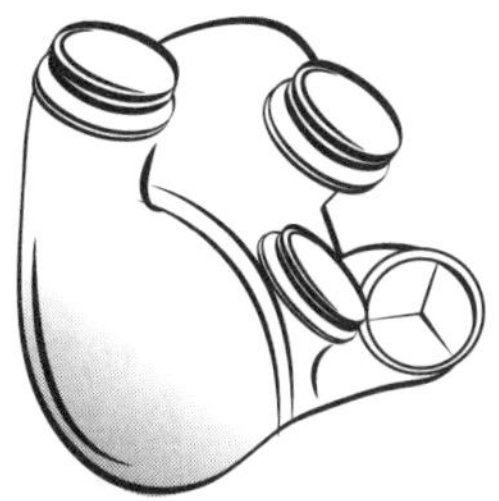

Purpose

The purpose of this assessment probe is to elicit students' ideas about the role of different professions in contributing to the solution of a problem that benefits society, such as the development of a lifesaving device. The probe is designed to determine if students realize that not only do engineers and scientists work closely, but they also work in teams that require the contributions of people in a variety of other professions, such as nurses, doctors, psychologists, and therapists.

Type of Probe

Justified list

Related Key Ideas

- Engineers create solutions for problems that deeply affect people, including some that save lives.
- Engineers nearly always work in teams, often including people from other professions.

Explanation

The best answer is all of the professions listed, and more than the ones listed are involved in various aspects of developing medical technology. Engineering is not a solitary process. It not only involves collaboration between engineers and scientists, but also involves contributions from a variety of professionals who bring different knowledge and experience to developing the solution. The artificial heart is an example of how several branches of science, engineering, mathematics, medicine, and manufacturing contributed to the development of a lifesaving device that affects society, especially since heart disease is the number one cause of death in the United States:

- *Cardiologists* are medical doctors who specialize in diagnosing and treating heart disease. They bring expertise in how hearts fail, and determine under what circumstances a replacement heart is needed.
- *Materials scientists* are needed to select materials for fabricating the artificial heart so that it is durable and compatible with the body's tissues.
- *Surgeons* are critical to advising on safely implanting an artificial heart into a patient.
- *Manufacturing engineers* work closely with design engineers to make sure the design

can be produced in large numbers safely, reliably, and economically.

- *Biochemists* provide expertise in how biological tissues and molecules interact with artificial heart materials so that the technology is safe and reliable for years of use.
- *Mechanical engineers* design the physical structure, pumping, and fluid flow required for an artificial heart to work safely for a long time.
- *Veterinarians* are needed for the animal trials, which must be performed before an artificial heart can be implanted into a human.
- *Intensive care nurses* provide experienced guidance on procedures for pre- and post-operation patient care for successful implant surgery and recovery.
- *Psychologists* can advise on appropriate methods for managing the worries and stresses that patients may have about artificial heart surgery and recovery.
- *Electrical engineers* are needed to develop the electronic power sources, controls, and remote communications systems of an artificial heart.
- *Biomedical engineers* have multidisciplinary expertise to help integrate the design across a range of factors such as biomaterials, structure, manufacturing, and long-term reliability.
- *Physical therapists* provide expertise on rehabilitation protocols specialized for an artificial heart that may have interacted differently from a real heart during exercise.
- *Mathematicians* have expertise in the advanced computational methods used in modeling the complex fluid flows in a heart, and of course all of the other professions listed have to use math as some part of their work.

Administering the Probe

This probe is best used with grades 3–12. With elementary students, we suggest modifying the probe to list a more limited set of professions likely to be familiar to the class and age level. In all cases, before using this probe, clarify each of the professions listed that students may not be familiar with. Show students actual pictures of an artificial heart. The probe can be extended by having students describe how they think each of the professions they selected contributed to the development of the artificial heart.

Connections to the Three Dimensions (NRC 2012; NGSS Lead States 2013)

- DCI: ETS2.A: Interdependence of Science, Engineering, and Technology
- Connection to the Nature of Science: Science Is a Human Endeavor

Related Research

- Foster and Ganesh (2013) evaluated a sixth-grade science and engineering unit in which students design and construct models for artificial hearts. Qualitative and quantitative data were collected from 32 students who participated in the unit. A preliminary analysis showed positive learning gains, ranging from 62% to 100% for student teams, in achieving three science learning objectives: the ability to define an engineering design process, develop a heart model, and engage in argument from evidence about the strengths and limitations of their models.
- Hmelo, Holton, and Kolodner (2000) tested a similar unit, but in this case the goal was to develop a model of an artificial lung. The researchers compared 42 sixth-grade students in two classrooms who engaged in the hands-on unit with

a comparison group from the same school who learned about the human respiratory system through direct instruction. All students received pre- and post-knowledge tests and 20 students in the experimental class were interviewed. Students engaged in engineering artificial lungs learned significantly more than those receiving direct instruction, and they learned to view the respiratory system more systematically. Due to the short duration of the activity, students learned more about structure than function and causal behaviors. According to the researchers:

> *"Before students began their design activity, they discussed what they might need to build their artificial lung. They offered suggestions that included such materials as broccoli, straws, balloons, and sponges. This was the beginning of a pattern in which students focused on designs that look like the lung rather than work like the lung. Many of them did, however, move on toward more function-oriented models." (Hmelo, Holton, and Kolodner al. 2000, p. 265)*

- Kelly et al. (2007) wrote about an afterschool program taught by university students to students in grades 6 and 7 at a middle school, with strong representation by girls and minorities. The treatment consisted of four modules, each requiring about 10 hours over five days to complete. Thirty-nine students in two afterschool classes were given a post-survey at the end of each module. Quantitative and qualitative analysis of the post-surveys were consistent with prior studies—that boys tended to enjoy computer programming and robotics more than girls did, and that girls rated a unit on biomedical engineering higher than boys did. The researchers concluded that activities that combine hands-on activities, role models, mentoring, and career exploration improve girls' self-confidence and interest in STEM courses and careers.

Suggestions for Instruction and Assessment

- Students of all ages are fascinated by the human body (especially their own!), and this probe is a great discussion starter for not only engaging students' interest in engineering, science, and health as careers, but also inspiring all students to take an interest in heart-healthy living.
- Have students discuss whether the development of the first artificial heart, the Jarvik 7, was a scientific advancement, an engineering advancement, or both.
- Have students research the development of recent artificial hearts and share how various professions, from the probe and any others they find, contribute to the development of the artificial heart.
- Brainstorm a class list of other devices for improving and saving lives that involve different types of engineers and other professions. The list will make for fun discussion and likely involve the families of all students. Here are some examples: eyeglasses, hearing aids, artificial knees/hips/shoulders, kidney dialysis machines, cardiac pacemakers, automatic insulation pumps, dentures, and tooth implants.
- Compare how designing the first artificial heart was similar to designing the first rocket that took humans to the Moon. What contributions from different engineering, scientific, medical, and other professions were similar? It is also meaningful to compare how the two efforts are different in methods, team size, goals, and impact on society.

References

Foster, C., and T. Ganesh. 2013. Engineering the human heart in the sixth grade classroom. Paper presented at the 2013 IEEE Frontiers in Education Conference, Oklahoma City. *https://ieeexplore.ieee.org/abstract/document/6685040.*

Hmelo, C. E., D. L. Holton, and J. L. Kolodner. 2000. Designing to learn about complex systems. *Journal of the Learning Sciences* 9 (3): 247–298.

Kelly, G., P. Klenk, G. Ybarra, and L. A. Cox. 2007. Assessment of gender differences on ratings of engineering learning modules in middle-school youth in an after-school setting. Paper presented at the American Society for Engineering Education's 2007 Annual Conference and Exposition, Honolulu. *https://peer.asee.org/2404.*

National Research Council (NRC). 2012. *A framework for K–12 science education: Practices, crosscutting concepts, and core ideas.* Washington, DC: National Academies Press.

NGSS Lead States. 2013. *Next Generation Science Standards: For states, by states.* Washington, DC: National Academies Press. *www.nextgenscience.org/next-generation-science-standards.*

How Are Science and Engineering Similar?

Science
Science and Engineering
Engineering

Put an X next to the things that describe **both** science and engineering.

___ **A.** Uses creativity and imagination

___ **B.** Includes people from all genders and cultures

___ **C.** People work in teams

___ **D.** Uses systematic investigations

___ **E.** Uses models and simulations

___ **F.** Uses math and computing

___ **G.** Identifies problems and designs solutions

___ **H.** Shares and presents ideas and information

___ **I.** Makes decisions based on evidence

___ **J.** Connects knowledge from different fields

___ **K.** Asks and answers questions about nature

___ **L.** Depends on people from other fields

___ **M.** Is based on curiosity and concern for our world

Explain your thinking about similarities and differences between science and engineering.

____

____

____

____

¿En Qué Se Parecen la Ciencia y la Ingeniería?

Ciencia

Ciencia e Ingeniería

Ingeniería

Pon una X al lado de las cosas que describen **tanto** la ciencia como la ingeniería.

___ **A.** Usa creatividad e imaginación

___ **B.** Incluye personas de todos los géneros y culturas

___ **C.** Las personas trabajan en equipos

___ **D.** Utiliza investigaciones sistemáticas

___ **E.** Usa modelos y simulaciones

___ **F.** Usa matemáticas y computación

___ **G.** Identifica problemas y diseña soluciones

___ **H.** Comparte y presenta ideas e información

___ **I.** Tomar decisiones basadas en evidencia

___ **J.** Conecta el conocimiento de diferentes campos

___ **K.** Hace y responde preguntas sobre la naturaleza

___ **L.** Depende de personas de otros campos

___ **M.** Basado en la curiosidad y preocupación por nuestro mundo

Explica tu pensamiento sobre las similitudes y diferencias entre la ciencia y la ingeniería.

____

____

____

____

____

How Are Science and Engineering Similar?

Teacher Notes

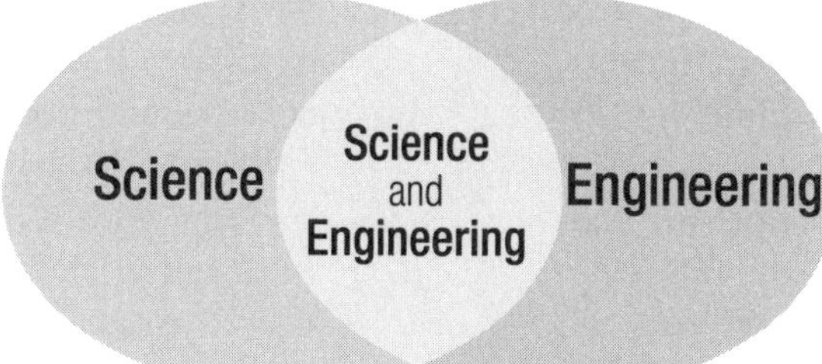

Purpose

The purpose of this assessment probe is to elicit students' ideas about the ways in which science and engineering are similar. The probe is designed to reveal whether students understand that science and engineering share many similarities including people, practices, and skills.

Type of Probe

Justified list

Related Key Ideas

- Engineers include people from many different ethnic and cultural backgrounds.
- Engineering as a profession is open to people with a wide variety of interests and capabilities.
- Engineering and science share many common practices.

Explanation

The best answer is A, B, C, D, E, F, H, I, J, L, and M—most of the items are common to both science and engineering. The only descriptions that do not describe both science and engineering are "G. Identifies problems and designs solutions" (this is primarily the work of engineering) and "K. Asks and answers questions about nature" (which is primarily the work of science). Although they have different purposes, with science focused on discovering new knowledge about the natural world, and engineering focused on improving our world, students are often unaware that there are many similarities between the fields. Many of the similarities in the probe are from the Science and Engineering Practices that form one of the three dimensions of the *NGSS* (*https://ngss.nsta.org/PracticesFull.aspx*). The other general category of similarities are behavioral and cultural aspects of the professions that are often associated with both engineering and science, including participation by people of all genders and cultures, the value of creativity and imagination, working in teams, and being motivated by curiosity and concern for our world.

Administering the Probe

This probe is best used with students in grades 5–12. The explanation can be extended by

having students explain how science and engineering support each other.

Connections to the Three Dimensions (NRC 2012; NGSS Lead States 2013)

- Science and Engineering Practices
  - Asking Questions (for science) and Defining Problems (for engineering)
  - Developing and Using Models
  - Planning and Carrying Out Investigations
  - Analyzing and Interpreting Data
  - Using Mathematics and Computational Thinking
  - Constructing Explanations (for science) and Designing Solutions (for engineering)
  - Engaging in Argument From Evidence
  - Obtaining, Evaluating, and Communicating Information
- Connection to the Nature of Science: Science Is a Human Endeavor

Related Research

- Fralick et al. (2009) asked approximately 1,600 middle school students to draw either a scientist or an engineer at work, and developed a checklist to code the results. Students involved in this study frequently perceived scientists as working indoors conducting experiments, and engineers as working outdoors in manual labor. However, a large fraction of the students had no perception of engineering.
- Rivale et. al. (2011) presented the Draw-an-Engineer test to 335 fifth-grade students. Only three boys in the entire sample drew a female engineer, suggesting that societal bias toward who belongs in the engineering profession is established early. Thirty-seven percent of the girls' drawings in which gender could be discerned included a female engineer. Girls were more likely to draw a female engineer if they were taught by a female graduate teaching fellow.

Suggestions for Instruction and Assessment

- This probe can be used with the card sort strategy. Print the descriptions on cards and have students sort them into descriptions of science, descriptions of engineering, and descriptions of both science and engineering.
- Following discussion of the probe, present students with a list of different fields of engineering and science and ask them to discuss how engineers and scientists in those fields engage in similar practices. Example questions include the following:
  - How do aerospace engineers who design airplanes and Earth scientists who study the formation of mountains and valleys use models?
  - How do automotive engineers who design cars and marine biologists who study fish populations use investigations and data analysis?
  - How do petroleum engineers who identify places where oil may be located and chemists who investigate the source of water pollution develop arguments based on evidence?
- Scientists and engineers often work together on projects. For example, some cities have identified problems with their water supply, such as finding high concentrations of lead in the water. Ask your students to consider what scientists would do to contribute to a solution for this problem. What role would the engineers play? Why is it important for the scientists and engineers to work together on projects like this?
- Have students identify similarities and differences between science and engineering practices to help them see that most

overlap. An excellent article for teachers is "Scientific and Engineering Practices in K–12 Classrooms" (Bybee 2011).

References

Bybee, R. W. 2011. Scientific and engineering practices in K–12 classrooms. *Science Scope* 35 (4): 6–11, 13.

Fralick, B., J. Kearn, S. Thompson, and J. Lyons. 2009. How middle schoolers draw engineers and scientists. *Journal of Science Education and Technology* 18 (1): 60–73.

National Research Council (NRC). 2012. *A framework for K–12 science education: Practices, crosscutting concepts, and core ideas.* Washington, DC: National Academies Press.

NGSS Lead States. 2013. *Next Generation Science Standards: For states, by states.* Washington, DC: National Academies Press. *www.nextgenscience.org/next-generation-science-standards.*

Rivale, S., J. Yowell, J. Aiken, S. Adhikary, D. Knight, and J. F. Sullivan. 2011. Elementary students' perceptions of engineers. Paper presented at 2011 ASEE Annual Conference & Exposition, Vancouver, BC. *https://peer.asee.org/17833.*

Is Engineering Creative?

Three students were trying out a new airbrush in their art class and wondered how airbrushes were invented. They each had different ideas about creativity in art and engineering:

Amber: Artists are creative. Engineers do not need to be creative.

Marisa: Both artists and engineers are creative.

Paulo: Engineers are creative when they are working on the parts of the product design that everyone can see. For the other parts, engineers don't need to be creative.

Who do you agree with the most? ________________ Explain why you agree.

____

____

____

____

____

¿La Ingeniería Es Creativa?

Tres estudiantes estaban usando un nuevo aerógrafo en su clase de arte. Se preguntaban cómo se inventaron los aerógrafos. Cada uno tenía ideas diferentes sobre la creatividad en el arte y la ingeniería:

Amber: Los artistas son creativos. Los ingenieros no necesitan ser creativos.

Marisa: Artistas e ingenieros son creativos.

Paulo: Los ingenieros son creativos cuando trabajan en las partes del diseño del producto que todos pueden ver. Para las otras partes, los ingenieros no necesitan ser creativos.

¿Con quién estás más de acuerdo? ________________ Explica por qué estás de acuerdo.

____

____

____

____

____

Is Engineering Creative?

Teacher Notes

Purpose

The purpose of this assessment probe is to elicit students' ideas about creativity in engineering. The probe is designed to determine the extent to which students think that creativity is an important part of engineering.

Type of Probe

Friendly talk

Related Key Idea

- Engineering is a highly creative process.

Explanation

The best answer is Marisa's: "Both artists and engineers are creative." Creativity involves the development of novel solutions in any field. While it is easy to see the creative aspect involved in art, it may be less obvious to see the creative aspect involved in engineering. Contrary to the common conception that engineering is just a straightforward, dry, methodical process, engineering also involves a lot of imagination, creativity, and innovation. To solve problems, engineers often think "outside the box." For example, photo IDs and passwords have long been a standard means for ensuring that only authorized individuals have access to a secure area, such as a bank vault. However, photo IDs can be forged and passwords stolen. A creative solution to this problem was the development of biometric technologies that can read a fingerprint or scan a person's iris. Also, while the development of products that must have an appealing visual design, such as a new car, is a creative aspect of engineering, so is creativity involved in developing processes and systems that we cannot see. One example is the software inside a computing system that organizes, searches, and calculates many different types of information; another is the precision mechanics inside a classic wristwatch.

Administering the Probe

This probe is best used in grades 3–12. You can extend the probe by asking students to describe an example to support their explanation.

Connection to the Three Dimensions (NRC 2012; NGSS Lead States 2013)

- Connection to the Nature of Science: Science Is a Human Endeavor

Related Research

- Rivale et al. (2011) administered the Draw-an-Engineer test to 335 fifth-grade students before and after engaging in engineering activities taught by engineering graduate students, and learning about key messages concerning engineering that the graduate students derived from a study by the National Academy of Engineering (*Changing the Conversation,* National Academy of Engineering 2008). These messages included the following:
    - Engineers make a world of difference.
    - Engineers are creative problem solvers.
    - Engineers help shape the future.
    - Engineering is essential to our health, happiness, and safety.

  Analysis of the results showed that the messages made a significant difference in students' knowledge of what engineers do.
- Frey and Powers (2012) reported on a reality-based television show called *Design Squad* in which competing teams of teenage students solved engineering design challenges to illustrate the excitement and enjoyment that comes from creative technical work. The theme of using a creative process to address challenges and learn from experimentation was a common thread through almost every episode. A summative assessment of the first season with 139 children who viewed four episodes indicated that the program positively influenced viewers' attitudes about engineering, including the ideas that engineers solve problems that affect real people, that engineers help make people's lives better, that engineering is NOT boring, and that men are NOT better at engineering than women.
- Rorrer, Sanders, and Knight (2005) engaged high school students in summer courses titled "The Science of Guitars" and "Musical Recording." These creative "hook" courses were followed by more rigorous "bridge" courses to prepare students for college engineering. Quantitative and qualitative methods were used to collect data from 26 high school students engaged in these courses. Content knowledge tests showed a significant gain between the pre- and posttests for the guitar course, and there was a significant gain in students' confidence in their knowledge about the pathway to college. Lack of student proficiency in math was the greatest challenge the instructors encountered.

Suggestions for Instruction and Assessment

- Follow up the probe with a discussion of what makes art a creative process and compare it to engineering as a creative process.
- When teaching the nature of science, expand it to include the nature of engineering. Several ideas about science as a human endeavor also apply to engineering. For example, in the *NGSS*, a grade 3–5 idea is that "creativity and imagination are important to science." Expand this to include engineering. Similarly, a high school idea in the *NGSS* states, "Scientific knowledge is a result of human endeavor, imagination, and creativity." Expand this to include engineering solutions (NGSS Lead States 2013).
- Have students come up with lists of things engineers have developed that demonstrate creative ways to solve a problem.
- Have students generate a list of problems in their everyday lives that could be solved

through engineering and brainstorm some creative ways to address these problems.

- When students are engaged in engineering challenges, explicitly point out the variety of different solutions students come up with and how this reflects creativity in their thinking.
- Have students view episodes of *Design Squad,* which can be streamed from the website *https://pbskids.org/designsquad.* This video series illustrates how young people can solve meaningful problems through creative engineering. Ask students to identify problems from their own daily lives that would be good projects for a team of students learning engineering.
- Have students choose a side and debate whether visual design (aesthetics) is part of engineering or a separate process.
- Have your students try these exercises to increase the number of creative ideas during brainstorming:
    - See who can come up with the wildest or craziest idea. Write the ideas down and see if any of them suggest something that might actually work.
    - Choose one of the ideas developed so far and think of variations that might make it better.
    - Choose two ideas brainstormed so far and combine them to come up with a different idea.
    - Have one student choose a word randomly from the dictionary. The team considers connections between the random word and the problem they are trying to solve. Do this for several words, and then share ideas.

References

Frey, D. D., and B. Powers. 2012. Designing *Design Squad:* Developing and assessing a children's television program about engineering. *Journal of Pre-College Engineering Education Research* 2 (1): 1–20. *http://docs.lib.purdue.edu/jpeer.*

National Academy of Engineering. 2008. *Changing the conversation: Messages for improving public understanding of engineering.* Washington, DC: National Academies Press. *https://doi.org/10.17226/12187.*

National Research Council (NRC). 2012. *A framework for K–12 science education: Practices, crosscutting concepts, and core ideas.* Washington, DC: National Academies Press.

NGSS Lead States. 2013. *Next Generation Science Standards: For states, by states.* Washington, DC: National Academies Press. *www.nextgenscience.org/next-generation-science-standards.*

Rivale, S., J. Yowell, J. Aiken, S. Adhikary, D. Knight, and J. F. Sullivan. 2011. Elementary students' perceptions of engineers. Paper presented at 2011 ASEE Annual Conference & Exposition, Vancouver, BC. *https://peer.asee.org/17833.*

Rorrer, R., R. Sanders, and D. Knight. 2005. Tapping hidden talent. Paper presented at ASEE's 2005 Annual Conference, Portland, OR. *https://peer.asee.org/15531.*

Reasons for Success

Two student teams were growing plants in indoor gardens of their own design. Students from the Blue Team saw that the Green Team's plants were larger than their plants. They are discussing how to re-engineer their garden design to improve plant growth. Each member of the Blue Team has a different idea:

Cynthia: They used some chemical tests to check the water to make sure it was OK for the kinds of plants they grew. Maybe we should have done that.

Marko: They used special growing lights. I didn't think it would matter, but I guess it did.

Stella: We can't be sure if it was the chemical tests or the grow lights, or something else, like the seeds they used or good air flow. We need to try some tests ourselves and discover what things cause the biggest improvement.

Who has the most useful way of thinking about the problem? ________________
Explain why you think that is the most useful way to come up with a solution.

____

____

____

____

____

Razones Para el Exito

Dos equipos de estudiantes cultivaban plantas en jardines interiores de su propio diseño. Los estudiantes del Equipo Azul vieron que las plantas del Equipo Verde eran más grandes que sus plantas. Están discutiendo cómo reingeñar el diseño de su jardín para mejorar el crecimiento de las plantas. Cada uno tiene una idea diferente:

Cynthia: Utilizaron algunas pruebas químicas para verificar el agua y asegurarse de que estaba bien para el tipo de plantas que cultivaban. Tal vez deberíamos haber hecho eso.

Marko: Usaron luces especiales de crecimiento. No pensé que importaría, pero supongo que sí.

Stella: No podemos estar seguros de si fueron las pruebas químicas o las luces de crecimiento, o algo más, como las semillas que usaron o el buen flujo de aire. Necesitamos probar algunas pruebas nosotros mismos y descubrir qué cosas causan la mejora mas grandé.

¿Quién tiene la forma más útil de pensar sobre el problema? ____________________
Explica por qué crees que esa es la forma más útil de encontrar una solución.

____

____

____

____

____

Reasons for Success

Teacher Notes

Purpose

The purpose of this assessment probe is to elicit students' ideas about the use of evidence and reasoning to support a claim. The probe is designed to see if students apply systematic methods of reasoning from evidence and controlled experimentation when they engineer a solution to a problem, which in this case is to promote greater growth of their plants.

Type of Probe

Friendly talk

Related Key Ideas

- Both engineering and science require that arguments be based on evidence and logical reasoning.
- Both scientists and engineers perform experiments that involve control and manipulation of variables.

Explanation

Stella has the most useful answer: "We can't be sure if it was the chemical tests or the grow lights, or something else, like the seeds they used or good air flow. We need to try some tests ourselves and discover what things cause the biggest improvement." Cynthia and Marko both have scientific knowledge and good ideas for what may explain the greater success of the other team. However, neither Marko's nor Cynthia's claims are fully supported by the available evidence. Stella correctly applies scientific thinking by recognizing that other factors could explain the other team's success, that more tests would be needed to gather enough evidence to determine which factors resulted in greater growth, and then that evidence would guide the re-engineering of their garden design for greater plant growth.

Administering the Probe

This probe is best used with students in grades 3–12. While it may be evident to students that Stella has the most useful answer, their explanations will reveal whether they recognize the role of scientific thinking in analyzing the problem. For upper elementary and middle school students, look for evidence that their explanations include the important idea that either the use of special grow lights, or the chemical tests performed by the other team,

or another factor could explain why the other group's garden was more successful. Middle and high school students may note that two or more factors may be responsible, and a series of experiments would be needed to acquire enough evidence to support a causal claim.

Connections to the Three Dimensions (NRC 2012; NGSS Lead States 2013)

- SEP: Engaging in Argument From Evidence
- SEP: Planning and Carrying Out Investigations
- CCC: Cause and Effect

Related Research

- Wright, Wendell, and Paugh (2018) carried out a case study of fourth- and fifth-grade students in an urban-like environment. Students participated in 18 hours of engineering design using the Engineering is Elementary curriculum. Each of the three units that students participated in required them to argue about the best solutions based on evidence. The researchers observed that the students were reluctant to engage in any argumentation, and instead agreed to combine their ideas rather than debate pros and cons. The researchers suspected that the students' reluctance to argue from evidence was a coping strategy in an environment where teachers discouraged any sort of conflict.
- Crismond and Adams (2012) reviewed hundreds of studies focused on students' abilities to engage in engineering design. One of their findings is that beginning students have little self-awareness of their design thinking: "Beginners do design tacitly when they act with little or no awareness of what they are doing, do not articulate what knowledge they know or need to know to further their investigations, and pay scant attention to the progress they make, obstacles they encounter, or design values that influence their decisions. They can fail to review steps they have taken or to examine the assumptions underlying their framing of the design problem" (p. 772). (In the probe, Stella is the only student who recognizes that her team should not make unwarranted assumptions, but rather that they should investigate the effect of the most likely variables before concluding why the other team's design was superior.)

Suggestions for Instruction and Assessment

- Lead a discussion about whether Marko and Cynthia's claims are supported by evidence. (There is some evidence in support of their claims but the evidence is weak because alternative explanations of the other group's success are possible.)
- Follow up this probe by having students describe how they would plan and carry out an investigation to find out why the Blue Team's garden design does not perform as well as the Green Team's garden.
- This probe uncovers a potential confusion between *correlation* (two events seem to be related) and *causation* (one event causes another). Ask students to think of situations where two events are correlated but one does not actually cause the other. For example, the number and severity of storms in the world are increasing, and glaciers are melting around the globe. One of these phenomena did not cause the other; both are likely caused by a third variable—climate change.
- Discuss the idea that causal reasoning is directional. For example, the number of extremely hot days in western states is increasing, and the number and severity of wildfires is also increasing in these states. The hot weather is an important causal factor in the increase in wildfires, nearly

all of which occur in summer months. However, the wildfires are not causing the hot weather.

- Emphasize that the practice of supporting claims with evidence and logical arguments applies to both science and engineering, even though goals may be different.

References

Crismond, D. P., and R. S. Adams. 2012. The informed design teaching and learning matrix. *Journal of Engineering Education* 101 (4): 738–797.

National Research Council (NRC). 2012. *A framework for K–12 science education: Practices, crosscutting concepts, and core ideas.* Washington, DC: National Academies Press.

NGSS Lead States. 2013. *Next Generation Science Standards: For states, by states.* Washington, DC: National Academies Press. *www.nextgenscience.org/next-generation-science-standards.*

Wright, C. G., K. B. Wendell, and P. P. Paugh. 2018. "Just put it together to make no commotion": Re-imagining urban elementary students' participation in engineering design practices. *International Journal of Education in Mathematics, Science and Technology* 6 (3): 285–301.

Is Engineering Experimental?

Jose and his mom are making biscuits for dinner. Jose encounters a problem with one of the ingredients:

Jose: Mom, we have a problem! I've mixed everything for the biscuits but we're out of baking powder. The recipe calls for one teaspoon of baking powder. Can I skip the baking powder? It's just one teaspoon.

Mom: No. The baking powder makes if fluffy. If you don't have baking powder you can use baking soda and lemon juice instead.

Jose: OK, but how much baking soda and lemon juice should I use?

Mom: I don't know. You'll just have to try an experiment to discover the best combination.

Jose: But Mom, cooks don't do experiments! I'll just take a guess and it will probably taste good enough.

Who has the best approach to the problem? ________________ Explain your thinking.

____

____

____

____

____

¿Es la Ingeniería Experimental?

José y su madre están haciendo galletas para la cena. Jose encuentra un problema con uno de los ingredientes:

Jose: Mamá, tenemos un problema! He mezclado todo para las galletas, pero no tenemos polvo para hornear. La receta requiere una cucharadita de polvo para hornear. ¿Puedo omitir el polvo de hornear? Es solo una cucharadita.

Mom: No. El polvo de hornear hace que sea esponjoso. Si no tienes polvo de hornear, puedes usar bicarbonato de sodio y jugo de limón.

Jose: Muy bien, pero ¿cuánto bicarbonato de sodio y jugo de limón debo usar?

Mom: No lo sé. Solo tendrá que probar un experimento para descubrir la mejor combinación.

Jose: Pero mamá, ¡los cocineros no hacen experimentos! Voy a adivinaré y probablemente sabrá lo suficientemente bueno.

¿Quién tiene el mejor enfoque para el problema? ____________________ Explica lo que piensas.

____

____

____

____

____

Is Engineering Experimental?

Teacher Notes

Purpose

The purpose of this assessment probe is to elicit students' ideas about using experimentation to solve a problem. The probe is designed to see if students recognize that (1) solving a problem in baking is considered engineering and (2) planned experimentation is something that is critical to engineering as well as science. The probe also provides a scenario to elicit students' ideas on how one might do an experiment in an everyday situation.

Type of Probe

Follow the dialogue

Related Key Ideas

- Engineering requires scientific and mathematical thinking.
- Both scientists and engineers perform experiments that involve control and manipulation of variables.

Explanation

Mom has the best approach to the problem. The goal of the probe is to see if students understand that experiments can be used for solving problems in science, engineering, and everyday life. This scenario is an engineering problem—to determine the amount of two ingredients that will produce fluffy biscuits. Jose's mom is correct that the best way to solve the problem is to conduct an experiment in which the value of all the variables is kept the same except for the experimental variable. It may be surprising to students that conducting an experiment to solve a problem in baking is engineering, but it is!

Administering the Probe

This probe is best used with students in grades 3–12. Students as young as third graders are ready to start developing the concept of a fair test before encountering the formal concept of controlled experiments. The probe can be extended by having students describe the approach that could be used to solve the problem, and looking carefully to see if students recognize the importance of controlling variables when testing a solution.

Connection to the Three Dimensions (NRC 2012; NGSS Lead States 2013)

- SEP: Planning and Carrying Out Investigations

Related Research

- Crismond and Adams' (2012) review of hundreds of studies of students' engineering design skills found the following: "Beginning designers run few or no tests on their design prototypes. When they do them, they conduct confounded experiments by changing multiple variables in a single experiment, which yields little understanding about potential solutions. Informed designers run valid tests as part of technological investigations that help them to learn quickly about design variables, users, and materials, to understand how things work, and to optimize the performance of the prototypes they decide to develop" (p. 765).
- Cunningham and Kelly (2017) took an ethnographic approach to examining how one experienced elementary teacher presented the Engineering is Elementary curriculum to 24 gifted and talented fourth graders. The study provided a rich description of how the teacher organized the class and guided students as they conducted controlled experiments and analyzed data to inform their engineering projects.

Suggestions for Instruction and Assessment

- Follow up by having students describe the approach either Jose's mother or Jose should use. The point is to understand their thinking about the need for controlled experimentation. Students may have an interest in understanding what are useful experiments in this real-life cooking situation, so a follow-up to this probe can have students help describe an experiment that Jose could do to determine how to make good biscuits without baking powder. For example, Jose could mix all of the ingredients, including a measured amount of lemon juice, but leave out the baking soda. He then could divide the dough into quarters, and combine each quarter with a different amount of baking soda. He would need to carefully label the biscuits before baking them. The biscuits that come out with the right amount of fluffiness would indicate the right ratio of lemon juice to baking soda. The main point is that the batches should be different in only one way, to find out the effect of the experimental variable (in this case, baking soda).
- Stress to students that engineering is a process of solving problems through technological design, from simple everyday problems to complex global problems. Cooking is a technological process, and your students may be surprised to learn that solving problems and developing new recipes in cooking is a kind of engineering. In fact, *food engineering* is a specialized profession concerned with food processing and safety.
- Provide other problem scenarios that students can use to practice identifying and controlling variables to test the design of a solution to a problem.
- *America's Test Kitchen* is a television show, magazine, and series of podcasts that use an engineering design process to develop the best recipes (*www.americastestkitchen.com*). Each recipe begins by identifying criteria for a successful result, and proceeds to test different ingredients and methods of cooking to meet the criteria. You can find some of the recipes on the *America's Test Kitchen* website with descriptions of the methods used to develop them. Be sure to have your students read the section included with the recipes titled "Why This

Recipe Works." Here's an example from one of the website's recipes:

> *To develop a sticky toffee pudding cake recipe with tolerable sweetness and a moist, tender crumb, we broke with tradition when choosing a sweetener, using brown sugar instead of treacle. To get the moist cakes we were looking for, we not only baked them in a water bath but also covered the pan with aluminum foil to seal in the steam while they cooked.*

- Food engineers developing new recipes often use taste tests in which different versions of a product are labeled in a way that does not give any information about how the versions differ (e.g., using the labels "A," "B," and "C"). This is called a "blind test," since the tasters have no information about the foods they are tasting. Ask your students what they think the purpose of a blind test is.
- To help students see that controlled experimentation helps with other kinds of recipes in non-food applications, discuss the idea that cooking is a kind of chemical engineering, in which different ingredients are combined and processed in various ways to achieve a product that is different from any of the ingredients. Paints and glues are examples of other products developed through chemical engineering. Explain the concept of chemical engineering to students and ask them to come up with other examples (such as cement, plastics, and metal alloys).

References

Crismond, D. P., and R. S. Adams. 2012. The informed design teaching and learning matrix. *Journal of Engineering Education* 101 (4): 738–797.

Cunningham, C. M., and G. J. Kelly. 2017. Framing engineering practices in elementary school classrooms. *International Journal of Engineering Education 33* (1): 295–307.

National Research Council (NRC). 2012. *A framework for K–12 science education: Practices, crosscutting concepts, and core ideas.* Washington, DC: National Academies Press.

NGSS Lead States. 2013. *Next Generation Science Standards: For states, by states.* Washington, DC: National Academies Press. *www.nextgenscience.org/next-generation-science-standards.*

Is It Rocket Science?

Three sisters watched a rocket launch at Kennedy Space Center for the new mission to Mars. They were really excited about the idea of going into space. They each want to get jobs someday designing, building, and launching rockets.

Sabrina: I think we need to study science to get a job in rocketry.

Ebony: I think we need to study engineering to get a job in rocketry.

Keisha: I think we can study science or engineering because science and engineering are both needed for developing rockets.

Who do you agree with the most? ________________ Explain your thinking.

____

____

____

____

____

¿Es Ciencia de Cohetes?

Tres hermanas vieron el lanzamiento de un cohete en Centro Espacial Kennedy para la nueva misión a Marte. Estaban realmente emocionados por ir al espacio. Cada uno de ellas quiere conseguir trabajo algún día diseñando, construyendo y lanzando cohetes.

Sabrina: Creo que necesitas estudiar ciencia para conseguir un trabajo en cohetería.

Ebony: Creo que necesitas estudiar ingeniería para conseguir un trabajo en cohetería.

Keisha: Creo que puedes estudiar ciencia o ingeniería porque la ciencia y la ingeniería son necesarias para desarrollar cohetes.

¿Con quién estás más de acuerdo?__________________ Explica lo que piensas.

____

____

____

____

____

Is It Rocket Science?

Teacher Notes

Purpose

The purpose of this assessment probe is to elicit students' ideas about the importance of both engineering and science for creating technology. The probe is designed to determine if students recognize that working in the field of rocketry requires both engineering and science.

Type of Probe

Friendly talk

Related Key Ideas

- Science and engineering complement each other and help each other progress.
- Scientists discover new knowledge, and engineers create solutions to meet people's needs.

Explanation

The best answer is Keisha's: "I think we can study science or engineering because science and engineering are both needed for developing rockets." Rocketry draws on a wide range of disciplines. Some example types of science and engineering and how they relate to rocketry are listed in the table that follows, and there are dozens more.

Engineering	Science
Chemical: fuels and propulsion systems	Physics: navigation
Mechanical: structural design of the spacecraft	Chemists: surface coatings and materials
Electrical: control and communications systems	Materials: almost anything in a rocket!
Biomedical: life support systems for health	Biologists: human physiology in space
Computer: electronics and software systems	Geologists: structure of planets and moons

Administering the Probe

This probe is best used with students in grades 6–12. The probe can be extended by asking students to describe the work that scientists and engineers do in rocketry.

Connections to the Three Dimensions (NRC 2012; NGSS Lead States 2013)

- DCI: ETS2.B. Influence of Science, Engineering, and Technology on Society and the Natural World
- Connection to the Nature of Science: Science Is a Human Endeavor

Related Research

- Many studies show that few students are aware of the variety of engineering careers and of what engineers do, and that interest in engineering careers is especially low among girls. High, Thomas, and Redmond (2010) compared the effects of three interventions for students in grades 6 and 7 to counter this trend. They compared 850 students who participated in Get a Grip summer camp, in which the final challenge was to design a prosthetic arm, with 550 students as controls who did not participate in the program. Students who participated in Get a Grip outperformed controls in their awareness of engineering and interest in engineering as a potential career. Among students who participated in the program, girls' confidence and belief in their own abilities and potential in science and math was significantly more positive than the boys' beliefs in the girls' abilities to do science and math. Also, the girls who participated in the summer camp demonstrated significantly more knowledge of what an engineer does and interest in pursuing an engineering career than those who did not attend the camp.
- McGrath et al. (2009) reported on the Build IT project, which aims to increase students' interest in engineering by constructing underwater robots using Lego Mindstorms. Data were collected from 22 middle school classes and 18 high school classes. Students were given pre-tests, post-tests, and attitude surveys. The Build IT curriculum increased students' awareness of engineering, their interest in engineering careers, and their understanding of an engineering design process and iterative design.

Suggestions for Instruction and Assessment

- As a follow up to this probe, start with a discussion about whether it is possible to design, build, and launch rockets only with the work of scientists, or only with the work of engineers. Or does it require both?
- Ask students to make a prediction about NASA's workforce. Does NASA employ more scientists or more engineers? NASA publishes a full description of their workforce on a website that it updates every two weeks. To support or refute their prediction, have students check NASA's employment website at *https://wicn.nssc.nasa.gov/wicn_cubes.html.* As of June 2019, NASA had 16,433 full-time employees. Of those, about 5,000 were in various management or clerical positions. Most of the other positions were in science and engineering, listed in the following categories.

Engineering	9,397
Physical Science	921
Mathematics	134
Biological Science	40
Social Science	17
Medical Science	1
Other	2
Total	10,512

The table shows that far more engineers than scientists are employed at NASA. Follow up with a discussion of what kinds of work engineers do at NASA and why NASA needs more engineers than scientists.

- Engage students in a discussion about how scientists contribute to the design, building, and launching of rockets. Which different fields of science might be involved? How would each contribute to rocket design, building, and launching?
- What types of engineers work on the design, building, and launch of rockets? What does each field of engineering contribute?
- A major cost of rocket launches has been single-use rockets. The Space Shuttle was designed to solve that problem with vehicles that could be reused many times. But the Space Shuttle is no longer in service. Have students research new rocket strategies for returning rockets to Earth for reuse. Have students view one of the spectacular YouTube videos that show rockets landing tail first on Earth and on barges in the ocean (e.g., *www.youtube.com/watch?v=lEr9cPpuAx8*). Discuss the kinds of science and engineering expertise needed to accomplish such a feat.

References

High, K., J. Thomas, and A. Redmond. 2010. Expanding middle school science and math learning: Measuring the effect of multiple engineering projects. Paper presented at the P–12 Engineering and Design Education Research Summit, Seaside, OR.

McGrath, E., S. Lowes, P. Lin, and J. Sayres. 2009. Analysis of middle and high school students' learning of science, mathematics and engineering concepts through a LEGO underwater robotics design challenge. Paper presented at ASEE's 2009 Annual Conference & Exposition, Austin, TX. *https://peer.asee.org/4794.*

National Research Council (NRC). 2012. *A framework for K–12 science education: Practices, crosscutting concepts, and core ideas.* Washington, DC: National Academies Press.

NGSS Lead States. 2013. *Next Generation Science Standards: For states, by states.* Washington, DC: National Academies Press. *www.nextgenscience.org/next-generation-science-standards.*

Section 3

Defining Problems

Key Ideas Matrix for Probes #17–#23

PROBES	# 17 An Engineering Design Process	#18 How Do Engineers Solve Problems?	#19 What's the Problem?	#20 Who Needs It?	#21 Is It an Engineering Problem?	#22 Criteria and Constraints	#23 Pizza Problem
GRADE-LEVEL USE →	**3–12**	**3–12**	**3–12**	**3–12**	**5–12**	**5–12**	**3–12**
RELATED KEY IDEAS ↓							
An engineering design process (EDP) is a systematic method for defining and solving problems.	X	X					
It is important to analyze a situation to determine the problem that needs to be solved.			X				
Identifying a "client" helps engineers be clear about whose needs the solution must meet.				X			
It is important to identify which problems are good candidates for engineering solutions and which are not.					X		
To define a problem, it is necessary to identify the criteria and constraints for a successful solution.						X	
Criteria are the features that define a successful solution.						X	
Constraints are the limits on the design process and solution.						X	
Research can be done many ways, including physical and digital searches, conducting scientific investigations, interviews with clients, and studying how similar problems were solved in the past.							X

Teaching and Learning Considerations

As pointed out by educational researcher David Crismond:

> *For decades, science curricula and teacher professional development materials have used the problem-solving context of design tasks to engage students in learning and applying big ideas of science, while making few or no references to engineering design. (Crismond 2013, p. 74)*

The situation that Crismond accurately described started to change with the publication of *A Framework for K–12 Science Education* (the *Framework*; NRC 2012) and eventual adoption of the *Next Generation Science Standards* (*NGSS*; NGSS Lead States 2013). Those documents called on students to understand an engineering design process (EDP) as a core idea, at the same level as Newton's laws of motion or the concept of genetic heredity, and to be able to apply an EDP to define and solve problems. The probes in Sections 3 and 4 are designed to elicit your students' current understanding of an EDP and how to apply it in real situations.

Beginning with an EDP, the probes in Section 3 are designed to reveal your students' understanding of an important skill that is valuable far beyond engineering: investigating a problem before beginning to solve it. Probes include the question of determining who cares about the solution, whether or not the problem can be solved through engineering, and how to define the problem in terms of criteria and constraints. It ends with a probe about the value of research to take advantage of prior successes and user feedback.

Further information about the attributes of design and engineering capabilities of students in grade bands K–2, 3–5, 6–8, and 9–12 is available in the *Standards for Technological Literacy* (ITEEA 2007) on pages 89–138.

References

Crismond, D. 2013. Troubleshooting: A bridge that connects engineering design and scientific inquiry. *Science Scope* 36 (6): 74–79.

International Technology and Engineering Educators Association (ITEEA). 2007. *Standards for Technological Literacy: Content for the study of technology.* 3rd ed. Reston, VA: ITEEA. *www.iteea.org/File.aspx?id=67767.*

National Research Council (NRC). 2012. *A framework for K–12 science education: Practices, crosscutting concepts, and core ideas.* Washington, DC: National Academies Press.

NGSS Lead States. 2013. *Next Generation Science Standards: For states, by states.* Washington, DC: National Academies Press. *www.nextgenscience.org/next-generation-science-standards.*

An Engineering Design Process

Engineering is a systematic method for solving problems. On the left side are several different parts of an engineering design process. Draw a line from each box on the left to the number on the right to show what you think might be a useful order of steps for solving a problem.

Improve the design by studying test results.	**1st**
Research the problem.	**2nd**
Define the problem.	**3rd**
Test the solution to see if it solves the problem.	**4th**
Brainstorm solutions.	**5th**
Choose the best idea to make a physical model (prototype).	**6th**
Communicate the final solution.	**7th**
Build a prototype for testing.	**8th**

Explain why you chose this sequence of steps.

____

____

____

____

____

El Proceso de Diseño de Ingeniería

La ingeniería es un método sistemático para resolver problemas. A la izquierda hay varias partes diferentes del proceso de diseño de ingeniería. Dibuja una línea desde cada cuadro a la izquierda hasta el número a la derecha para mostrar la orden que crees que es útil para resolver problemas.

Mejore el diseño, estudiando los resultados de las pruebas.	**1.º**
Investigue el problema.	**2.º**
Define el problema.	**3.º**
Pruebe la solución para ver si resuelve el problema.	**4.º**
Genera ideas.	**5.º**
Elija la mejor idea para hacer un modelo físico (prototipo).	**6.º**
Comunicar la solución final.	**7.º**
Crea un prototipo para probar.	**8.º**

Explica por qué elegiste esta secuencia de pasos.

____

____

____

____

____

An Engineering Design Process

Teacher Notes

Improve the design by studying test results	1st
Research the problem	2nd
Define the problem	3rd
Test the solution to see if it solves the problem	4th
Brainstorm solutions	5th
Choose the best idea to make a physical model (prototype)	6th
Communicate the final solution	7th
Build a prototype for testing	8th

Purpose

The purpose of this assessment probe is to elicit students' ideas about an engineering design process. The probe is designed to show how students think of engineering design as a logical sequence.

Type of Probe

Sequencing

Related Key Idea

- An engineering design process (EDP) is a systematic method for defining and solving problems.

Explanation

There is no single best answer that puts all the steps in a correct order. The sequence students select and their explanation should show how they think about what they would do first, second, and so on when solving a problem, and that they realize that a systematic approach is really just a way of thinking logically and carefully—not rushing to the first solution that comes to mind or believing that there is a fixed set of steps that must always be followed in a definite sequence. The *NGSS* development team recognized that a wide variety of interpretations of an EDP are acceptable; so rather than choosing one vision of the process over another, the *NGSS* included a diagram of an EDP (below) that consists of three phases: define the problem, develop solutions, and optimize (NGSS Lead States 2013). These three broad phases encompass many versions of an EDP associated with various sets of standards and curricula.

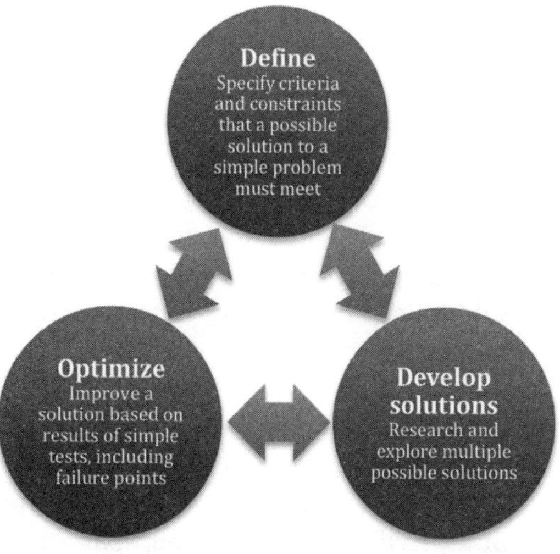

Administering the Probe

This probe is best used with students in grades 3–12. The probe can be used as a card sort. Provide pairs or small groups of students with a set of cards, with each part of the design process listed on a card. Students sort the cards, putting them into a logical sequence and explaining why they put them in that order.

Connections to the Three Dimensions (NRC 2012; NGSS Lead States 2013)

In the *NGSS,* an EDP is both a disciplinary core idea (DCI) and a science and engineering practice (SEP). That means students should be able to both describe an EDP and use it to define and solve problems.

There are three broad DCIs for understanding an EDP in grades K–12:

- ETS1.A: Defining and Delimiting Engineering Problems
- ETS1.B: Developing Possible Solutions
- ETS1.C: Optimizing the Design Solution

The K–12 SEPs often found to be useful in an EDP are as follows:

- Asking Questions (for science) and Defining Problems (for engineering)
- Developing and Using Models
- Planning and Carrying Out Investigations
- Analyzing and Interpreting Data
- Using Mathematics and Computational Thinking
- Constructing Explanations (for science) and Designing Solutions (for engineering)
- Engaging in Argument From Evidence
- Obtaining, Evaluating, and Communicating Information

Related Research

- Alemdar et al. (2017) developed an assessment of middle school students' understanding of an EDP that used 18 multiple-choice questions. The purpose of the assessment was to identify the extent of common misconceptions, such as the idea that an EDP is a step-by-step fixed process rather than an iterative process, in which students continually revise and improve prior designs. The researchers tested their assessment instrument with 44 middle school students by comparing the results of the multiple-choice assessment with individual interviews, and refined the instrument based on their findings. The researchers commented that students were more familiar with some phases of an EDP than with others. For example, they were better at evaluating existing designs than they were at initially creating designs, possibly because evaluation of designs was emphasized in their engineering curriculum.

Suggestions for Instruction and Assessment

- This probe can be used as a starting point to assess and discuss students' initial ideas about an EDP. The subsequent probes in this book are intended to assess students' understanding of each phase of an EDP, and to deepen the sophistication with which they consider and solve problems.
- There is a rich and vibrant literature on the history and nature of engineering design, including the popular writings of Henry Petroski, such as *Invention by Design: How Engineers Get From Thought to Thing* (Petroski 1996). Excerpts from these books can be used with students to provide them with a real-life, engaging glimpse into engineering design.
- For students in grades 6–12 (and possibly also grades 3–5 if vocabulary is made an early part of the discussion), the following diagram is a simple and dynamic view of an EDP that abstracts the key features of

many EDP models and uses the typical terminology used by engineering practitioners.

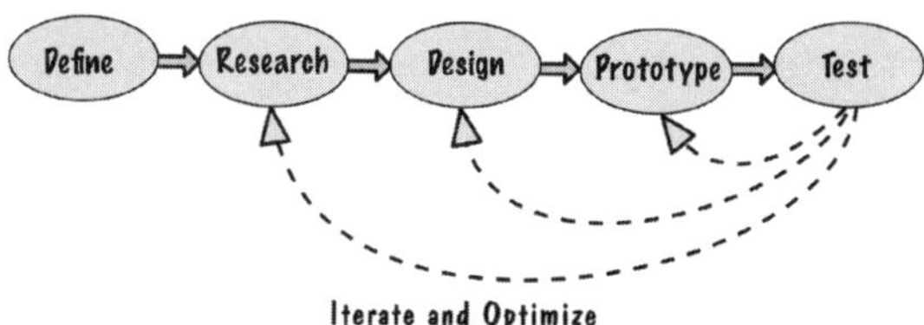

- **Define:** The first step involves developing a clear statement of the problem in terms of criteria for success, and constraints or limits of possible solutions. A unique aspect of engineering is the presence of a *client*—an individual or group of people who have a problem or need, and who can help identify the criteria and constraints.
- **Research:** Research should include exploring how similar problems have been solved in the past, and may include other kinds of studies, such as marketing, to see who would use the solution and what additional criteria or constraints should be considered.
- **Design:** This phase usually begins by considering many different possible solutions, synthesizing various ideas, and determining which idea is most likely to solve the problem. This phase also usually includes creating a model of the solution, which may be as simple as a paper sketch, a detailed two- or three-dimensional model using computer-aided design, or a full mathematical computer simulation model.
- **Prototype:** A prototype is a physical or software model that can be tested to determine how well the design meets the criteria and constraints of the problem. It can also be shown to potential users to see if they agree it meets their needs. Early prototypes can also be "virtual prototypes" that are detailed functional simulation models.
- **Test and Revise:** In many cases the first attempt will not completely solve the problem, leading to further research. The dotted arrows indicate that further research may or may not be needed. If improvements are obvious, the designers might construct and test a new prototype right away. If not, they may need to return to the design phase and think of new solutions. Or, if the test shows they are on the wrong track, they may need to conduct further research to learn more about what will solve the problem. This iterative process continues until the problem is solved. Once it is solved, students can continue to optimize the solution—that is, to improve it as much as possible.
- **Communicate:** Although not shown in the diagram, the final stage of an EDP is often a presentation describing the new design with a summary of the process so that the audience understands how the design was arrived at and why it is the best way to solve the problem.

- For K–5 grades, you may wish to consider a diagram with fewer steps and more daily language, such as the diagram of an EDP developed for use in the Engineering is Elementary curriculum by the Museum of Science in Boston. See *www.eie.org/overview/engineering-design-process.*
- A different way to think about an EDP for 6–12 students was developed as part of the *Design Squad* television program (Wolsky 2015). A clickable version of *Design Squad*'s EDP diagram can be found

online at *https://pbskids.org/designsquad/parentseducators/workshop/process.html.*

- The Massachusetts Department of Elementary and Secondary Education has included an EDP in its standards since 2001. Each subsequent improvement of the standards has included a modified diagram of an EDP. The most recent version of the state's EDP diagram and its relationship to science inquiry can be found in the *Massachusetts Science and Technology/Engineering Framework* (2016, pp. 98–100). The document can be downloaded at *www.doe.mass.edu/frameworks/scitech/2016-04.pdf.*
- Regardless of which EDP representation you use with your students, it is important to emphasize that engineering design is both logical and creative, and uses a systematic and iterative process. There is no firmly set official starting point or end point. Depending on the problem and what has been done so far, you can begin at any step, move back and forth between steps, or even repeat the entire cycle. The important point is that it is methodical and involves following a sequence of steps.

References

Alemdar, M., J. A. Lingle, S. A. Wind, and R. A. Moore. 2017. Developing an engineering design process assessment using think-aloud interviews. *International Journal for Engineering Education* 33 (1): 441–452.

National Research Council (NRC). 2012. *A framework for K–12 science education: Practices, crosscutting concepts, and core ideas.* Washington, DC: National Academies Press.

NGSS Lead States. 2013. *Next Generation Science Standards: For states, by states.* Washington, DC: National Academies Press. *www.nextgenscience.org/next-generation-science-standards.*

Petroski, H. 1996. *Invention by design: How engineers get from thought to thing.* Cambridge, MA: Harvard University Press.

Massachusetts Department of Elementary and Secondary Education. 2016. *Massachusetts science and technology/engineering curriculum framework. www.doe.mass.edu/frameworks/scitech/2016-04.pdf.*

Wolsky, M. 2015. Design Squad: Inspiring a new generation of engineers. In *The go-to guide to engineering curricula, grades 6–8,* ed. C. I. Sneider, 19–31. Thousand Oaks, CA: Corwin Press.

How Do Engineers Solve Problems?

Four students were having a discussion about how engineers solve problems. They each had a different idea about how engineers do their work:

Simone: I think engineers try one thing and if it doesn't work they try something else until they solve the problem.

Andrew: I think engineers have to be very organized so there are steps that every engineer has to follow in order when solving a problem.

Kazem: I think engineers follow steps in different orders for solving different problems.

Honaka: I think engineers use different methods but they always involve building things.

Which student do you agree with the most? ________________ Explain why you agree.

____

____

____

____

____

¿Cómo Resuelven Problemas los Ingenieros?

Cuatro estudiantes estaban discutiendo sobre cómo los ingenieros resuelven los problemas. Cada uno tenía una idea diferente sobre cómo los ingenieros hacen su trabajo:

Simone: Creo que los ingenieros prueban una cosa y, si no funciona, intentan otra hasta que resuelven el problema.

Andrew: Creo que los ingenieros deben ser muy organizados, por lo que hay pasos que cada ingeniero debe seguir para resolver un problema.

Kazem: Creo que los ingenieros seguen los pasos en diferentes órdenes para resolver diferentes problemas.

Honaka: Creo que los ingenieros utilizan diferentes procesos, pero siempre implican construir cosas.

¿Con qué estudiante estás más de acuerdo? ____________________ Explica por qué estás de acuerdo.

____

____

____

____

____

How Do Engineers Solve Problems?

Teacher Notes

Purpose

The purpose of this assessment probe is to elicit students' ideas about how engineers use an engineering design process to solve problems. The probe includes some common misconceptions about engineering design, and is designed to reveal whether students recognize that different approaches are used, depending on the problem.

Type of Probe

Friendly talk

Related Key Idea

- An engineering design process (EDP) is a systematic method for defining and solving problems.

Explanation

The best answer is Kazem's: "I think engineers follow steps in different orders for solving different problems." It is important for students to recognize that there is no one definite series of steps that engineers always follow. Although an EDP provides an overall framework, the order and total number of steps differ depending on the problem they are trying to solve.

Administering the Probe

This probe is best used with students in grades 3–12. Listen carefully to your students' explanations for their choices and encourage discussion. Extend the probe by asking students to explain why they do not agree with the other statements. Each of the alternative responses is a common misconceptions about engineering. For example, Simone's response suggests that engineering is a random trial and guessing process, rather than logical and systematic; Andrew's response suggests that engineers follow a rigid set of steps; and Honaka's response is indicative of the common idea that engineers just build things.

Connections to the Three Dimensions (NRC 2012; NGSS Lead States 2013)

In the *NGSS*, an EDP is both a disciplinary core idea (DCI) and a science and engineering practice (SEP). That means students should both be able to both describe an EDP and use it to define and solve problems.

There are three broad DCIs for understanding an EDP in grades K–12:

- ETS1.A: Defining and Delimiting Engineering Problems
- ETS1.B: Developing Possible Solutions
- ETS1.C: Optimizing the Design Solution

The K–12 SEPs often found to be useful in an EDP are as follows:

- Asking Questions (for science) and Defining Problems (for engineering)
- Developing and Using Models
- Planning and Carrying Out Investigations
- Analyzing and Interpreting Data
- Using Mathematics and Computational Thinking
- Constructing Explanations (for science) and Designing Solutions (for engineering)
- Engaging in Argument From Evidence
- Obtaining, Evaluating, and Communicating Information

Related Research

- Fralick et al. (2009) administered the Draw-an-Engineer test and Draw-a-Scientist test to approximately 1,600 middle school students, and developed a checklist to code the results. Students' perceptions of what engineers do was inferred from the actions of the people shown in their drawings: 31% were shown making something with their hands, 11% operating or driving machines, 10% designing, 2% explaining, 2% experimenting, and 2% observing. No action was depicted in 27% of the drawings. In 15%, no person was depicted at all. Overall, students involved in this study most frequently perceived scientists as working indoors conducting experiments, and engineers as working outdoors in manual labor.
- Hirsch et al. (2007) developed a survey instrument to measure middle school students' attitudes about engineering and knowledge of engineering careers. The instrument was then used to compare students who were exposed to engineering through math and science classes with those who were not. Students exposed to engineering were able to name more engineering professions, and were better able to give examples of what engineers do. However, no significant differences were found regarding their attitudes toward STEM.
- The idea that there is a common series of steps that is followed by all scientists is likely to be the most common myth of science (McComas 1998). Results of this probe for engineering may mirror the research that shows students think scientists follow a rigid, step-by-step method, called "the scientific method."

Suggestions for Instruction and Assessment

- This probe can be used as a starting point to assess and discuss students' initial ideas about how an EDP is used to solve actual problems. The subsequent probes in this book are intended to assess students' understanding of each phase of an EDP, and to deepen the sophistication with which they consider and solve problems.
- Be careful that the step-by-step portrayal of an EDP doesn't give the false impression that there is one set of steps that all engineers follow, and that the steps are followed in a definite order. This misconception is similar to students' view of "the scientific method" as indicating that science works only in a fixed serial fashion.
- When referring to engineering design, one approach to broaden students' mindsets is to use the indefinite article "an," rather than the definite article "the," since the

phrase "*the* engineering design process" implies that there is one way to approach an engineering problem. In contrast, "*An* engineering design process" implies that there is more than one way to solve a problem.

- Explain that even though engineers may take a different approach to every problem, their approach is always thoughtful and creative, as well as systematic. That is, they need to be sure they understand the problem that is to be solved, and they imagine and consider a variety of different solutions before deciding on the best one to test. Usually several tests are needed before engineers finalize the design, and they can always go back to a previous step.
- A similar probe, "Doing Science" in *Uncovering Student Ideas in Science, Volume 3,* reveals similar misunderstandings students have about the process and methods scientists use to do their work (Keeley, Eberle, and Dorsey 2008).
- Find out if any of your students have family or friends who are engineers and invite them to come to your class to describe their work. Spend time in advance of the visit encouraging your students to think of questions, such as the following: "What was your most interesting project?" "Do you always use the same method to solve a problem?" "Do the methods you use depend on the problem you are solving?"
- Demonstrate the idea that different problems lead to different EDPs by choosing two or three different problems and showing how each one would lead to a different starting point and different steps to reach a successful solution.
- Search the internet for short videos about what engineers do and how they solve problems. Select two or three to show your students.

References

Fralick, B., J. Kearn, S. Thompson, and J. Lyons. 2009. How middle schoolers draw engineers and scientists. *Journal of Science Education and Technology* 18 (1): 60–73.

Hirsch, L. S., J. D. Carpinelli, H. Kimmel, R. Rockland, and J. Bloom. 2007. The differential effects of pre-engineering curricula on middle school students' attitudes to and knowledge of engineering careers. Paper presented at the ASEE/IEEE 37th Frontiers in Education Conference, Milwaukee. *https://ieeexplore.ieee.org/stamp/stamp.jsp?arnumber=4417918.*

Keeley, P., F. Eberle, and C. Dorsey. 2008. *Uncovering student ideas in science, volume 3: Another 25 formative assessment probes.* Arlington, VA: NSTA Press.

McComas, W. F. 1998. The principal elements of the nature of science: Dispelling the myths. In *The nature of science in science education: Rationales and strategies,* 53–70. Boston: Kluwer Academic Publishers.

National Research Council (NRC). 2012. *A framework for K–12 science education: Practices, crosscutting concepts, and core ideas.* Washington, DC: National Academies Press.

NGSS Lead States. 2013. *Next Generation Science Standards: For states, by states.* Washington, DC: National Academies Press. *www.nextgenscience.org/next-generation-science-standards.*

What's the Problem?

Daniella and her friend Tyson are about to leave for school. Tyson finds out his bicycle has a flat tire. He's worried he will be late for school. Daniella and Tyson have different ideas about what is the most important problem to solve first:

Tyson: We need to figure out how to fix my tire.

Daniella: We need to figure out how to get to school on time.

Who identified the problem that should be solved first? ________________ Explain your thinking.

____

____

____

____

____

¿Cuál Es el Problema?

Daniella y su amigo Tyson se están preparando para ir a la escuela. Tyson nota que su bicicleta tiene una rueda pinchada. Le preocupa llegar tarde a la escuela. Tyson y Daniella tienen ideas diferentes sobre cuál es el problema más importante que deben resolver primero:

Tyson: Necesitamos descifrar cómo arreglar mi llanta.

Daniella: Necesitamos descifrar cómo llegar a la escuela a tiempo.

¿Quién identificó el problema que se debería resolver primero? ____________________
Explica lo que piensas.

____

____

____

____

____

What's the Problem?

Teacher Notes

Purpose

The purpose of this assessment probe is to elicit students' initial response to an ill-defined problematic situation. The probe is designed to reveal whether students recognize that in a problem situation, there is sometimes a more important problem that needs to be solved first.

Type of Probe

Opposing views

Related Key Idea

- It is important to analyze a situation to determine the problem that needs to be solved.

Explanation

The best answer is Daniella's: "We need to figure out how to get to school on time." It is likely that many students will focus on the image of the bicycle, and immediately assume that the problem is the flat tire. However, the more important and immediate problem to be solved is how to get to school on time. Stopping to repair the bicycle will likely make them even later. This kind of situation can also happen with seemingly straightforward engineering problems in which a problem situation needs to be more clearly defined. Sometimes an important or immediate problem needs to be addressed first, before the more obvious problem can be solved.

Administering the Probe

This probe is best used with grades 3–12. The probe can be extended by having students describe what they would do to solve the problem.

Connections to the Three Dimensions (NRC 2012; NGSS Lead States 2013)

- DCI: ETS1.A. Defining and Delimiting Engineering Problems
- SEP: Asking Questions and Defining Problems

Related Research

- Crismond and Adams' (2012) review of several hundred studies of how students solve problems found that "beginning designers feel that understanding the design

challenge is straightforward, and a matter of comprehending the basic task and its requirements. By perceiving the design task as a well-structured problem and believing there is a single correct answer, they can act prematurely and attempt to solve it immediately. Informed designers start by trying to learn as much as they can about the problem and delaying design decisions until they understand the problem fully. They set out to learn through research, brainstorming, and doing technological investigations of what the critical issues are in order to frame the problem effectively. They will later return to assess this framing after attempting to solve the challenge in case they need to modify the problem definition" (p. 747).

- Watkins, Spencer, and Hammer (2014) had fourth graders use fictional texts as a basis for identifying, scoping, and designing solutions for an engineering problem that the characters face. In contrast to Crismond and Adams' findings, the investigators found that the students did not treat design problems as well-defined straightforward tasks. Rather, they demonstrated promising beginnings of the ability to define a problem.

Suggestions for Instruction and Assessment

- Lead an all-class discussion as a follow-up to this probe. It is likely that students will disagree about what needs to be done. In that case, encourage more discussion.
- If all students believe the tire needs to be fixed, you can guide students to consider the other problem—getting to school on time. For example, you can ask students what happens when they are late for school, especially if they have been late several times before. How would their experience with being late for school inform how they would approach this problem?
- If most students say that the tire should be fixed first, ask them if they ever fixed a bicycle tire or watched someone do it. What is involved in fixing the tire? How long might it take to fix the tire? Then ask if they still think Tyson and Daniella could fix the tire and still get to school on time.
- Have the students share information about how they come to school in the morning. Do they ride the bus? Does a parent take them? Do they walk or ride a bicycle to school? What would they do if their usual means for getting to school failed? How would they solve the problem of getting to school?
- Have students come up with their own examples of a problem that at first seemed straightforward and easy to solve, but, on further examination, there were underlying problems that needed to be addressed first.
- Ask students to compare how this problem situation might be similar to problems that engineers encounter. Encourage them to guess about a situation, as any answer is a good start.
- Once students have completed this probe, have them discuss the problems involved in this situation. Initially, they may think that fixing a tire is an engineering problem and getting to school is not. However, a more accurate description is the reverse. Fixing tires is simply a step-by-step mechanical task that does not require design skills. However, solving how to get to school on time could be an engineering problem as students will need to define the problem, brainstorm multiple solutions, compare them, and select the best solution.

References

Crismond, D. P., and R. S. Adams. 2012. The informed design teaching and learning matrix. *Journal of Engineering Education* 101 (4): 738–797.

National Research Council (NRC). 2012. *A framework for K–12 science education: Practices, crosscutting concepts, and core ideas.* Washington, DC: National Academies Press.

NGSS Lead States. 2013. *Next Generation Science Standards: For states, by states.* Washington, DC: National Academies Press. *www.nextgenscience.org/next-generation-science-standards.*

Watkins, J., K. Spencer, and D. Hammer. 2014. Examining young students' problem scoping in engineering design. *Journal of Pre-College Engineering Education Research* 4 (1): 43–53.

Who Needs It?

Simone recently graduated from college with an engineering degree. She has just been hired by a company to design its new line of birdhouses. She knows that every product has a *client*. The *client* is usually a person or group who has a problem or need that requires a solution. Identifying clients is an important step early in an engineering design process. She asks her friends to help her identify a client for the birdhouses.

Ling: The client is the person who hired you. Just ask your employer to tell you as much as they can about what they want the new birdhouses to be like.

Annapurna: I think the client is the person who is likely to buy a birdhouse. If you meet that person's needs, then your employer will be happy. Go to a garden shop where they sell birdhouses and ask the customers what they are looking for.

Deepali: We need to think of this from the user's point of view. Your client is clearly the bird that will be living in the birdhouse. Go visit some gardens and parks to see which birdhouses attract the most birds!

Melvin: I think all three of you identified a client for the birdhouse.

Katrina: I disagree with all of you. The client is someone else.

Who do you agree with the most? ________________ Explain your thinking.

____

____

____

____

____

¿Quién lo Necesita?

Simone se graduó recientemente de la universidad con un título de ingeniería. Acaba de ser contratada por una empresa para diseñar su nueva línea de casas para pájaros. Ella sabe que cada producto tiene un *cliente*. El *cliente* es regularmente una persona o grupo que tiene un problema que requiere una solución. Identificar clientes es un paso inicial importante en un proceso de diseño de ingeniería. Ella le pide a sus amigos que la ayuden a identificar un cliente para las casas de pájaros.

Ling: El cliente es la persona que lo contrató. Solo pídale a su empleador que le diga todo lo que pueda sobre cómo quieren que sean las nuevas casas para pájaros.

Annapurna: Creo que el cliente es la persona que probablemente compre una casa para pájaros. Si satisface las necesidades de esa persona, su empleador estará muy contento. Vaya a una tienda donde venden pajareras y pregunte a los clientes qué están buscando.

Deepali: Tenemos que pensar en esto desde la vista del usuario. Su cliente es claramente el pájaro que vivirá en la casa para pájaros. ¡Visite algunos jardines y parques para ver qué pajareras atraen más aves!

Melvin: Creo que los tres identificaron un cliente para la casa para pájaros.

Katrina: No estoy de acuerdo con todos ustedes. El cliente es otra persona.

¿Con quién estás más de acuerdo? ____________________ Explica lo que piensas.

Who Needs It?

Teacher Notes

Purpose

The purpose of this assessment probe is to elicit students' ideas about who needs to be consulted when solving a problem (the client). The probe is designed to reveal the extent to which students recognize that the needs of a number of individuals (in this case an animal as well) have to be understood when defining a problem.

Type of Probe

Friendly talk

Related Key Idea

- Identifying a "client" helps engineers be clear about whose needs the solution must meet.

Explanation

The best answer is Melvin's: "I think all three of you identified a client for the birdhouse." As part of the process of defining a problem, engineers identify the *client*, which is the individual, company, organization, or other entity that has a need or a problem to be solved that uses the expertise of an engineer. It is important to be clear about whose needs will be met by the solution, and who else may be affected when the problem is solved. The birdhouse problem is especially interesting because, in addition to the person who first identified the need (the boss), and the person who will make a decision about whether or not to purchase it (the customer), there is a third client whose needs must also be taken into account (the bird). In engineering, the bird's position is sometimes referred to as "the end user." Melvin is correct that defining this problem will involve learning about the needs of all three "clients."

Administering the Probe

This probe is best used with students in grades 3–12. Be sure students first understand what is meant by the "client" in an engineering situation, and how its meaning differs from the word *customer.* For older students, in addition to *client,* discuss terms like *customer, buyer,* and *end user*—these can all be different people. Have them think about a situation in which their parents are the *buyers* but they are the *users.*

Connections to the Three Dimensions (NRC 2012; NGSS Lead States 2013)

- DCI: ETS1.A. Defining and Delimiting Engineering Problems
- DCI: ETS2.B. Influence of Science, Technology, and Engineering on Society and the Natural World
- SEP: Asking Questions and Defining Problems

Related Research

- Several researchers have investigated the value of using fictional characters in literature to spark discussion about engineering design. Portsmore, Watkins, and McCormick (2012) and Watkins, Spencer, and Hammer (2014) reported on a study of 24 fourth-grade students who identified problems the main characters faced in a fictional story. Pairs of students selected a problem to work on and designed solutions to help the characters by engaging in several types of planning and drawing activities.
- A similar teaching method is described by McCormick and Hammer (2016), who reported on a case study of two students during a fourth-grade engineering activity in which students framed problems based on fictional stories that provided "clients." The outcomes were encouraging but mixed, in part owing to variations in how students framed the task. While in some cases students focused on clients' needs, in other cases they focused on what they thought the teacher wanted.

Suggestions for Instruction and Assessment

- Help students gain insight into the distinction between cases in which the buyer of a technology is different from the end user. For example, ask them to think about situations in which parents are the buyers and the children are the users, or where someone is buying a gift for a friend who will be the end user. How should a person who is designing the technology decide on the qualities of a good solution? What if they please the buyer but not the end user? Or the end user but not the buyer?
- Identify an object that many of your students use every day, such as a backpack, a smartphone, or a common article of clothing. Ask the students who they thought the object was originally designed for, what the users might have said if a design engineer had asked them how they would use the object, and what characteristics they thought it should have and should not have.
- Ask students to imagine they have been hired by a school supplies company and asked to redesign an everyday object typically used in school, such as a cafeteria tray or locker. Who would they talk to in order to find out how the task should be defined? What questions should they ask the person?
- Is the best design always a good choice? What if the best design costs more money than the competition? How might an employer decide if the design that one of its engineers came up with will actually make money for the company?

References

McCormick, M. E., and D. Hammer. 2016. Stable beginnings in engineering design. *Journal of Pre-College Engineering Education Research* 6 (1): 45–54.

National Research Council (NRC). 2012. *A framework for K–12 science education: Practices, crosscutting concepts, and core ideas.* Washington, DC: National Academies Press.

NGSS Lead States. 2013. *Next Generation Science Standards: For states, by states.* Washington, DC: National Academies Press. *www.nextgenscience.org/next-generation-science-standards.*

Portsmore, M., J. Watkins, and M. McCormick. 2012. INSPIRE Engineering Education Summit, Washington, DC.

Watkins, J., K. Spencer, and D. Hammer. 2014. Examining young students' problem scoping in engineering design. *Journal of Pre-College Engineering Education Research* 4 (1): 43–53.

Is It an Engineering Problem?

People encounter all sorts of problems every day. Some can be solved effectively with an engineering design process (EDP); others require a different approach.

Can these problems be solved by an EDP?

Problem Situation	Yes	No	Maybe
1. Earbud cords get easily tangled when placed in a pocket.			
2. Cafeteria trays are slippery, so when students take their trays to a table, they often spill their drinks.			
3. A student has to decide whether to put money from her summer job into a bank account paying interest, or buy a bicycle to ride to an after-school job.			
4. Plants for a science project need to be watered every other day, but school will be closed for a week.			
5. When students get box lunches for a field trip, they waste food by throwing out items they don't like.			
6. A local food bank doesn't have enough food to feed all of the families that need help.			
7. A group of people wants to change the name of a school, but other people want to keep it the same.			
8. On rainy days, students track mud into the school.			
9. Inspectors found dangerous levels of lead in drinking water from many of the city's elementary schools.			
10. Elm trees in communities all over the country are dying from Dutch elm disease.			

Explain your thinking. How did you decide whether a problem is best solved using an EDP?

¿Es un Problema de Ingeniería?

La gente encuentra problemas de todo tipos todos los días. Algunos pueden resolverse de manera mucho más efectiva con el proceso de diseño de ingeniería; pero otros requieren un enfoque diferente.

¿Se pueden resolver estos problemas con el proceso de diseño de ingeniería?

Situación Problemática	Si	No	Tal Vez
1. Los cables de los auriculares se enredan fácilmente cuando se colocan en un bolsillo.			
2. Las bandejas de la cafetería son resbaladizas, por lo que cuando los estudiantes llevan sus bandejas a una mesa, a menudo derraman sus bebidas.			
3. Un estudiante tiene que decidir si es mejor poner dinero de su trabajo de verano en una cuenta bancaria que paga intereses, o comprar una bicicleta para ir al trabajo después de la escuela.			
4. Las plantas para un proyecto de ciencias deben regarse cada dos días, pero la escuela permanecerá cerrada por una semana.			
5. Cuando los estudiantes reciben almuerzos para una excursión, desperdician alimentos tirando artículos que no les gustan.			
6. Un banco de alimentos local no tiene suficientes alimentos para alimentar a todas las familias que necesitan ayuda.			
7. Un grupo de personas quiere cambiar el nombre de una escuela, pero otras personas quieren mantenerlo igual.			
8. En días lluviosos, los estudiantes traen lodo a la escuela.			
9. Los inspectores encontraron niveles peligrosos de plomo en el agua potable de muchas de las escuelas primarias de la ciudad.			
10. Los olmos están muriendo de la enfermedad holandesa del olmo.			

Explica lo que piensas. ¿Cómo decidistes si el problema se resuelve mejor utilizando el proceso de diseño de ingeniería?

Is It an Engineering Problem?

Teacher Notes

Problem Situation	Yes	No	Maybe
1. Earbud cords get easily tangled when placed in a pocket.			
2. Cafeteria trays are slippery, so when students take their trays to a table, they often spill their drinks.			
3. A student has to decide whether to put money from her summer job into a bank account paying interest, or buy a bicycle to ride to an after-school job.			
4. Plants for a science project need to be watered every other day, but school will be closed for a week.			
5. When students get box lunches for a field trip, they waste food by throwing out items they don't like.			
6. A local food bank doesn't have enough food to feed all of the families that need help.			
7. A group of people want to change the name of a school, but other people want to keep it the same.			
8. On rainy days, students track mud into the school.			
9. Inspectors found dangerous levels of lead in drinking water from many of the city's elementary schools.			
10. Elm trees in communities all over the country are dying from Dutch elm disease.			

Purpose

The purpose of this assessment probe is to elicit students' ideas about how to approach a problem. The probe is designed to gain insight into how students think about an engineering design process (EDP) and the kinds of problems that it can be used to solve.

Type of Probe

Justified list

Related Key Idea

- It is important to identify which problems are good candidates for engineering solutions and which are not.

Explanation

There are no absolute right or wrong answers to these questions. An EDP can be applied to any of them. However, some are much more likely to be addressed with an EDP than others. Here are some thoughts about each of these situations. Can these problems be solved by an EDP?

Problem Situation	Possible Answers
1. Earbud cords get easily tangled when placed in a pocket.	Yes. A device could be designed to hold the cord so it does not tangle.
2. Cafeteria trays are slippery, so when students take their trays to a table, they often spill their drinks.	Yes. A different kind of tray can be designed to hold the drinks better. Also, the current trays can be treated so they are non-skid or easier to hold firmly.
3. A student has to decide whether to put money from her summer job into a bank account paying interest, or buy a bicycle to ride to an after-school job.	Maybe. The student could use a systematic process to consider the need for a bicycle, or use math to see which decision gives her more money. However, instead of using these engineering processes, the decision could be based on how much the student wants a bicycle.

Continued

(*continued*)

Problem Situation	Possible Answers
4. Plants for a science project need to be watered every other day, but school will be closed for a week.	Yes. A device could be designed to provide water to the plants every other day automatically.
5. When students get box lunches for a field trip, they waste food by throwing out items they don't like.	Yes. Different solutions are possible, such as students picking the foods for their own box lunches. Each solution has different economic and social benefits that can be calculated.
6. A local food bank doesn't have enough food to feed all of the families that need help.	Yes. There may be other sources of food, such as unused food from nearby schools, or other types of food that are available at lower costs in bulk. An EDP approach would need to calculate the trade-offs between positive benefits such as lower costs and any negative impacts such as lower nutritional content.
7. A group of people wants to change the name of a school, but other people want to keep it the same.	Maybe. The cost of new signs or other changes could be taken into account when weighing different options systematically. However, such decisions are usually made through discussion and democratic votes.
8. On rainy days, students track mud into the school.	Yes. Once the source of the mud is known, different solutions could be proposed and tested, with each solution having different benefit/cost trade-offs.
9. Inspectors found dangerous levels of lead in drinking water from many of the city's elementary schools.	Yes. Inspectors would need to track down the source of the pollution, evaluate different options for cleaning or replacing the equipment containing lead, and select the solution with the best benefits and lowest costs for the city.
10. Elm trees in communities all over the country are dying from Dutch elm disease.	Maybe. Scientists and engineers can work together to determine how the disease is transmitted, seek and evaluate possible remedies, and recommend various methods with different benefits and costs so that people can choose suitable options for their community trees.

Administering the Probe

This probe is best used with students in grades 5–12. The best way to use this probe is to have individual students check off their answers first and then form pairs to discuss their ideas. Following the pair discussion, conduct a class vote on each problem, asking volunteers who voted for different answers to share their reasoning.

Connections to the Three Dimensions (NRC 2012; NGSS Lead States 2013)

- DCI: ETS1.A. Designing and Delimiting Engineering Problems
- SEP: Asking Questions and Defining Problems

Related Research

- By asking students which kinds of problems can be solved with an EDP, we are also learning what students think engineering is all about. English, Hudson, and Dawes (2011) studied the results of a short engineering lesson in Australia in which students discussed engineers they know, watched a DVD about engineering, interacted with websites about engineering, and researched famous engineers. After these experiences, students expressed the idea that engineers are creative, future-oriented, and artistic problem finders and solvers; planners and designers; seekers and inventors; and builders. The students also described them as adventurous, decisive, community-minded, reliable, and smart. Although students displayed a broader awareness of engineering than existing research suggests, there was limited knowledge of engineering fields and still a strong perception of engineering as largely involved in construction.

Suggestions for Instruction and Assessment

- Following the class vote and discussion (as described in the Administering the Probe section), have students summarize in one sentence when they think an EDP is best used in a problem situation. Guide students toward understanding that it is best used whenever a systematic approach would be helpful.
- The probe can be extended by having students describe how an EDP could be used for each of the problems they felt could be best solved systematically. Do not share the sample answers in the Explanation section until students have had an opportunity to explain their own ideas about how an EDP is applied to each problem situation.
- Have students extend the list by generating their own examples of problems to solve. Students could exchange their new problem ideas with others who would respond with yes, no, and maybe when it comes to deciding whether the problem could be solved by engineering. Have them justify their reasons for the yes, no, or maybe approaches.
- Have students identify a problem they encountered in real life in which an EDP was not used to solve the problem, but the problem could have been solved more effectively with an EDP.
- Ask students to name a few problems for which an EDP would obviously *not be a good approach* (such as naming a baby, or deciding what flavor of ice cream to order). Make a list of these on the board. Then do the same with problems for which an EDP *is clearly a good choice.* For each list, ask the students to describe, in one sentence, what all of the problems on the list have in common. Record these ideas. Finally, have the students explain, in their own words, when an EDP is and is not a good method for solving a problem.

References

English, L. D., P. B. Hudson, and L. A. Dawes. 2011. Middle school students' perceptions of engineering. STEM in Education Conference: Science, Technology, Engineering and Mathematics in Education Conference.

National Research Council (NRC). 2012. *A framework for K–12 science education: Practices, crosscutting concepts, and core ideas.* Washington, DC: National Academies Press.

NGSS Lead States. 2013. *Next Generation Science Standards: For states, by states.* Washington, DC: National Academies Press. *www.nextgenscience.org/next-generation-science-standards.*

Criteria and Constraints

Sophia decided to make a skateboard in her woodworking class. She bought the wheels, but had to shape the wood provided in class using hand tools. She wanted the skateboard to be light and strong and shaped just right with the nose and tail angled upward. Her first skateboard competition was in only six days.

Before she started designing, Sophia wrote down her definition of the problem—to make a skateboard that will be both easy to ride and that can "pop" off the ground. She then listed the criteria and constraints of the problem.

> ***Criteria*** are the features that define a successful solution.
>
> ***Constraints*** are the limits on the design process.

How many criteria and constraints can you identify in the description of the problem? Circle the number you find for each one:

Criteria:	1	2	3	4 or more
Constraints:	1	2	3	4 or more

Explain your thinking. List the things from the description that are considered criteria and the things that are constraints.

___

___

___

___

___

Criterios y Limitaciones

Sophia decidió hacer una patineta en su clase de carpintería. Compró las ruedas, pero tuvo que dar forma a la madera proporcionada en clase con herramientas manuales. Ella quería que fuera ligero y fuerte y que tuviera la forma correcta con la nariz y la cola en ángulo hacia arriba. Su primera patineta competencia va ser en solo seis días.

Antes de comenzar a diseñar, Sophia escribió su definición del problema: hacer una patineta que sea fácil de montar y «saltar» del suelo. Luego enumeró los criterios y las limitaciones del problema.

Los criterios son las características que definen una solución exitosa.

Las limitaciones son los límites en el proceso de diseño.

¿Cuántos criterios y limitaciones puedes identificar en la descripción del problema? Encierra en un círculo el número que puedes encontrar para cada uno:

Criterios: 1 2 3 4 o más

Limitaciones: 1 2 3 4 o más

Explica lo que piensas. Enumere las cosas de la descripción que se consideran criterios y las cosas que son limitaciones.

___

___

___

___

___

Criteria and Constraints

Teacher Notes

Purpose

The purpose of this assessment probe is to elicit students' ideas about criteria and constraints. The probe is designed to identify how students determine the criteria and constraints of a problem and whether one is more difficult to identify than the other.

Type of Probe

Quantifying probe

Related Key Ideas

- To define a problem, it is necessary to identify the criteria and constraints for a successful solution.
- Criteria are the features that define a successful solution.
- Constraints are the limits on the design process and solution.

Explanation

The best answer from the text description is 4 or more criteria and 4 or more constraints. Identifying as many as possible leads to better understanding of the problem. The probe defines the problem quite clearly, but the criteria and constraints are embedded in the text that explains what Sophia is attempting to do. The criteria and constraints embedded in the text description are as follows:

Criteria	Constraints
• Light • Strong • Nose angled up • Tail angled up • Easy to ride	• Use purchased wheels • Use wood given in class • Use hand tools • Be finished in six days

Most students have little difficulty identifying criteria, since the problem situation usually dictates the qualities of a good solution. However, students can have difficulty with identifying constraints. In truth there is often a "gray area" in which it is not easy to determine if a requirement is a criterion or a constraint. One way to think about this is that a constraint is a limitation or condition that must be satisfied by a design. Some examples of constraints are the project budget, schedule, or staffing available for the team. A criterion

is a required attribute of a design that can be specified and measured. For example, some criteria for a new car could be the fuel efficiency of the car, or the volume of cargo space in the back of the car.

Administering the Probe

This probe is best used with students in grades 5–12. Remind students to use the text description in the probe when selecting their answer. The probe can be extended to ask students to list other things, not described in the text, that could be criteria or constraints.

Connections to the Three Dimensions (NRC 2012; NGSS Lead States 2013)

- DCI: ETS1.A. Defining and Delimiting Engineering Problems
- SEP: Asking Questions and Defining Problems

Related Research

- Crismond and Adams' (2012) review of several hundred studies of how students solve problems found that "beginning designers ignore or pay too little attention to design criteria and constraints, and focus only on positive or negative aspects of their design ideas without thinking of associated benefits and trade-offs. Informed designers balance systems of benefits and trade-offs when they consider various plans, make design decisions, and justify them" (p. 761).
- Berland, Steingut, and Ko (2014) examined the impact of a high school course—UTeach, Engineer Your World—by administering 179 questionnaires and interviewing 16 students. The study found that students better understand and value those aspects of engineering design that are more qualitative (e.g., interviewing users, generating multiple solutions) than the more quantitative aspects of design in which students integrate traditional math and science into their designs.

Suggestions for Instruction and Assessment

- Have students note how Sophia's definition of the problem makes it easy to test her skateboard design (easy to maneuver when doing ollies and easy to "pop" off the ground).
- Extend the probe by having students describe how Sophia would test her skateboard. Each test should allow her to determine if her skateboard has met one or more of the criteria.
- Identify and describe a variety of situations in which an engineering design process can be used to solve a problem. Have students practice generating criteria and constraints that need to be considered before designing the solution.
- Choose a product that many students use every day, such as a pencil, pen, or backpack. Ask them to imagine that they are designing an improved version of the product. Assign them to work in pairs to list as many criteria as they can think of that need to be considered for a new and improved version. Ask the students to also think of at least two constraints that may limit the possible designs or development process.
- Select a sport that many students are familiar with, such as soccer or baseball, and ask students to imagine that some students were recently injured while playing that sport. A group of students and teachers were given the job of changing the rules for the sport. What are some criteria and constraints that they should consider when thinking up new rules?

References

Berland, L., R. Steingut, and P. Ko. 2014. High school student perceptions of the utility of the engineering design process: Creating opportunities to engage in engineering practices and apply math and science content. *Journal of Science Education Technology* 23 (6): 705–720.

Crismond, D. P., and R. S. Adams. 2012. The informed design teaching and learning matrix. *Journal of Engineering Education* 101 (4): 738–797.

National Research Council (NRC). 2012. *A framework for K–12 science education: Practices, crosscutting concepts, and core ideas.* Washington, DC: National Academies Press.

NGSS Lead States. 2013. *Next Generation Science Standards: For states, by states.* Washington, DC: National Academies Press. *www.nextgenscience.org/next-generation-science-standards.*

Pizza Problem

Business has been slow at Claire's Pizza Place. Claire thinks a better pizza will increase sales. She asks for help from a team of students from an engineering class at a local school. The team volunteers its time helping businesses with design projects. The student team visits Claire to learn about the pizza design project. After the meeting, the students have a discussion to share different ideas on improving the pizza and attracting more customers:

Meena: I know what we should do. Let's put chocolate and other kinds of candy on the pizzas. People like candy and they like pizza. That combination will be a winner!

Cybil: Let's design a new kind of sauce. That is what gives pizza its taste.

Deven: We should talk with as many people as possible to find out what they like and don't like about pizza.

Asami: Let's see what the other pizza places in town are doing. We can find out what kind of crust, sauce, and toppings people like best, and how much those pizza places charge.

Who do you think has the best idea? ________________ Explain why you think it is the best idea.

____

____

____

____

____

Problema de Pizza

El negocio ha sido lento en la pizzería de Claire. Ella piensa que una mejor pizza aumentará las ventas. Ella pide ayuda a un equipo de estudiantes de una clase de ingeniería en una escuela local. El equipo ofrece su tiempo como voluntario para ayudar a las empresas con proyectos de diseño. El equipo de estudiantes visita a Claire para aprender sobre el proyecto de diseño de pizza. Después de la reunión, tienen una discusión para compartir diferentes ideas sobre cómo mejorar las pizzas y atraer más clientes:

Meena: Sé lo que debemos hacer. Pongamos chocolate y otros tipos de dulces en las pizzas. A la gente le gustan los dulces y les gusta la pizza. ¡Esa combinación será un ganador!

Cybil: Diseñemos un nuevo tipo de salsa. Eso es lo que le da sabor a la pizza.

Deven: Deberíamos hablar con tanta gente como sea posible para saber qué les gusta y qué no les gusta de la pizza.

Asami: Veamos qué están haciendo las otras pizzerías de la ciudad. Podemos averiguar qué tipo de corteza, salsa, y coberturas prefieren las personas y cuánto cobran las pizzas.

¿Quién crees que tiene la mejor idea? ____________________ Explica por qué crees que es la mejor idea.

____

____

____

____

____

Pizza Problem

Teacher Notes

Purpose

The purpose of this assessment probe is to elicit students' ideas about researching a problem. The probe is designed to determine the extent to which students are aware of the need to learn as much as possible about how others have solved a problem, and what the potential users want, before brainstorming possible solutions.

Type of Probe

Friendly talk

Related Key Idea

- Research can be done many ways, including physical and digital searches, conducting scientific investigations, interviews with clients, and studying how similar problems were solved in the past.

Explanation

The best idea is Deven's: "We should talk with as many people as possible to find out what they like and don't like about pizza." Talking with potential customers will help the employees identify ideas as well as criteria and constraints for good solutions. Asami also has a good idea. Perhaps the other pizza restaurants in town already solved the problem of what people like best about pizza, so all that's necessary is to find out why they are successful. However, the other pizza restaurants may not have found the best possible solutions, so it is important to learn more about what different people like and do not like about pizza. While Meena's enthusiasm is admirable and Cybil's idea might improve their existing pizzas, neither is the best approach. First, without attempting to learn why their pizza is selling poorly, the team may be trying to solve the wrong problem. Second, Meena may become so enthralled by her first idea that she is unwilling to consider other ideas, even if they are more promising.

Administering the Probe

This probe is best used with students in grades 3–12. This probe can be extended by asking students to explain how the problem with the pizza is similar to problems engineers face in their work.

Connections to the Three Dimensions (NRC 2012; NGSS Lead States 2013)

- DCI: ETS1.A. Defining and Delimiting Engineering Problems
- DCI: ETS1.B. Developing Possible Solutions

Related Research

- Crismond and Adams' (2012) review of several hundred studies of how students solve problems found that "beginning designers skip doing research in favor of generating solutions immediately. Informed designers instead do research on users, write product histories, and collect information on manufacturing methods, materials, and product standards to build understandings of the problem and potential solutions" (p. 752).

Suggestions for Instruction and Assessment

- After discussing the best answer to the probe, ask students to think about what they like or do not like about pizza. Write their ideas on the board, under the topics "what we like about pizza" and "what we don't like about pizza." Ask the students to imagine they are developing ideas for a pizza restaurant that would appeal to as many of their classmates as possible. Which ideas would enable the restaurant to be most successful?
- Have students come up with a list of interview questions to ask potential customers, which could then be used to inform their potential solution to improving pizza sales.
- Have students select a product that needs improvement and suggest ways to conduct research that might improve the product.
- An important point is to ask students to think about how the pizza problem relates to the work of engineers. If students have difficulty coming up with ideas, ask them about what the pizza problem suggests about automotive engineers who must come up with new and better model cars every year.
- In some cases what people *want* is not the only, or even the most important, information that engineers need to define the problems they are working on. Consider the process of designing an airplane. What do aviation engineers need to consider that has little to do with what air travelers want?

References

Crismond, D. P., and R. S. Adams. 2012. The informed design teaching and learning matrix. *Journal of Engineering Education* 101 (4): 738–797.

National Research Council (NRC). 2012. *A framework for K–12 science education: Practices, crosscutting concepts, and core ideas.* Washington, DC: National Academies Press.

NGSS Lead States. 2013. *Next Generation Science Standards: For states, by states.* Washington, DC: National Academies Press. *www.nextgenscience.org/next-generation-science-standards.*

Section 4

Designing and Testing Solutions

Key Ideas Matrix for Probes #24–#32

PROBES	#24 Brainstorming	#25 Engineering and Nature	#26 Is It Affordable?	#27 What Is a Product's Life Cycle?	#28 Engineers' Models	#29 Picking the Best Solution	#30 Designing With Math and Science	#31 Testing for Success	#32 Making It Better
GRADE-LEVEL USE →	**3–12**	**3–12**	**5–12**	**3–12**	**3–12**	**6–12**	**6–12**	**3–12**	**3–12**
RELATED KEY IDEAS ↓									
Rules for brainstorming help problem solvers generate a wide variety of ideas.	X								
Teamwork is essential for effective problem solving.	X								
Nature can provide inspiration and ideas for solutions.		X							
Making a solution more affordable requires trade-offs between cost and other features.			X						
Life cycle analysis includes environmental impacts of all stages of production from raw material to disposal.				X					
A decision matrix is one way to systematically compare solutions to see which is best.						X			
Trade-offs are decisions that involve reducing some desirable features in favor of others.			X			X			
Most engineering problems require the application of math and science.							X		
Engineers make and use various kinds of models at different stages of a design process.					X			X	
Solutions must be tested to see if they meet the criteria and constraints of the problem.								X	
A prototype is a model that can be tested to check if the design solves the problem.								X	
Optimization involves further tests and improvements to find the best possible solution.									X
Many different methods can be used to optimize a design.									X

Teaching and Learning Considerations

The probes in this section are intended to help you refine instruction to address your students' current thinking about an engineering design process. It starts with Probe 24, "Brainstorming," which is a process of generating as many different ideas as possible. Research studies show that beginning designers often become fixated on one idea too early in the brainstorming process, and fail to develop a wide variety of solutions. This probe will reveal if brainstorming is a process you will need to spend time on.

Extending the idea that brainstorming needs to result in a variety of ideas, Probe 25, "Engineering and Nature," is a challenge using bio-inspired design to come up with some creative ideas. Probes 26 and 27 are about important concerns that need to be taken into account early in a design process—that a successful design needs to be affordable and have minimal negative impacts on the environment. Probes 28–30 are about methods used by engineers to understand and solve problems—modeling, choosing the solution that is most likely to solve the problem, and using math and science in engineering. Probe 31 is about the important process of testing to be certain a proposed design will solve the problem, and Probe 32 concerns the process of arriving at the best solution (optimization) through repeated testing and improvement (iteration).

Key ideas about technology and engineering related to modern technological systems, including medical, agricultural, energy and power, information and communication, transportation, manufacturing, and construction technologies, can be found in the *Standards for Technological Literacy* (ITEEA 2007) on pages 139–219.

Reference

International Technology and Engineering Educators Association (ITEEA). 2007. *Standards for Technological Literacy: Content for the study of technology.* 3rd ed. Reston, VA: ITEEA. *www.iteea.org/File.aspx?id=67767.*

Brainstorming

The student council has been asked to brainstorm ideas for keeping the school grounds free of trash. After discussing the problem, they came up with one idea. Here's part of their discussion:

Lee: We should make the younger students do the clean-up.

Loretta: That's not fair! Just because you're older doesn't mean that you get to order the younger students around.

Sidney: I think Lee has an interesting idea. If the younger students have to spend a lot of time cleaning up, they'll learn their lesson early to keep the school clean.

Shirley: Just the opposite! Once they get to our grade, they'll feel they don't have to clean anymore and they will litter even more!

Is this a good example of brainstorming? (circle your answer): Yes No Maybe

Explain your thinking.

____

____

____

____

____

Sesión de Reunión Creativa

Se le ha pedido al consejo estudiantil que comparta ideas para mantener los terrenos escolares libres de basura. Después de discutir el problema, se les ocurrió una idea. Aquí está parte de su discusión:

Lee: Deberíamos hacer que los estudiantes más jóvenes limpien el patio de la escuela.

Loretta: ¡Eso no es justo! El hecho de que seas mayor no significa que puedas decirles a los estudiantes más jóvenes qué hacer.

Sidney: Creo que Lee tiene una idea interesante. Si los estudiantes más jóvenes tienen que pasar mucho tiempo limpiando, aprenderán su lección temprano para mantener limpia la escuela.

Shirley: ¡Todo lo contrario! Una vez que lleguen a nuestro grado, sentirán que ya no tienen que limpiar y ¡arrojarán más basura!

¿Es este un buen ejemplo de una sesión de reunión creativa? (Encierre en un círculo tu respuesta): Sí No Tal vez

Explica lo que piensas.

____

____

____

____

____

Brainstorming

Teacher Notes

Purpose

The purpose of this assessment probe is to elicit students' ideas about brainstorming. The probe is designed to determine the extent to which students understand the norms of good teamwork during the brainstorming process.

Type of Probe

Follow the dialogue

Related Key Ideas

- Rules for brainstorming help problem solvers generate a wide variety of ideas.
- Teamwork is essential for effective problem solving.

Explanation

The best answer is "No." Once a problem has been defined and researched, the next phase is to generate as many creative ideas as possible. Guidelines for brainstorming ideas typically focus on ways to encourage lots of different ideas and avoid discussions about just one or two ideas. For example, one common guideline is not to evaluate or criticize ideas during brainstorming. Another is to write down everything, including ideas that are not practical, since they may lead to other ideas that are practical. The scenario in this probe illustrates how violating rules for brainstorming stops the creative brainstorming process and results in a very small number of ideas.

Administering the Probe

This probe is best used with students in grades 3–12. It can be used to introduce the process of brainstorming or can be used after students have been introduced to brainstorming to determine how well they understand the norms for brainstorming.

Connection to the Three Dimensions (NRC 2012; NGSS Lead States 2013)

- DCI: ETS1.B. Developing Possible Solutions

Related Research

- Crismond and Adams' (2012) review of several hundred studies of how students solve problems found a considerable body of research findings. Following are quotes

from their research review (Crismond and Adams 2012, pp. 755–757):

> *Beginning designers can start their design work with very few or even just one idea, which they may not want to discard, add to, or revise. Informed designers want to design with an abundance of ideas and practice idea fluency using techniques such as brainstorming and divergent thinking to explore the design space and at least initially seek to avoid favoring any single solution.*
>
> *Brainstorming, one of the hallmark strategies of designers, involves generating a wide-ranging collection of ideas while deliberately withholding criticisms and deferring judgment on the quality of those ideas. Ideas may be generated using a wide range of materials and modes of expression such as sketching, which is linked with the re-interpretation of design ideas. Making analogies can encourage grouping and connecting ideas in unexpected ways, which may enhance ideation and lead to the development of new products.*
>
> *Research has suggested that instructions emphasizing the withholding of criticism are less effective than stressing the creation of a large numbers of ideas in producing more ideas and more good ideas. However, simply asking students to generate lots of ideas with or without judging them ignores the nontrivial challenge of developing an ability to brainstorm. Without scaffolding, elementary and middle school student designers often do not generate multiple solutions when facing design problems. Attempts to mandate idea fluency have backfired on teachers, especially when such edicts are issued without explaining to students the reasons for such a dictate. McCormick, Murphy, and Davidson (1994) describe cases where teachers required students to include three or four candidate ideas in their design portfolios, from which students purportedly would select one idea for implementation. The authors tell how students actually generated the requisite alternative solutions after completing their design projects that were based on a single idea. They referred to such student portfolio work as a "veneer of accomplishment." When, without rationale, brainstorming is proposed for students to do, they can treat it as a required classroom ritual that they perform superficially, if at all.*

Suggestions for Instruction and Assessment

- Choose an engineering context and have the class work in small groups of five to six students to practice brainstorming.
- Explain the purpose of brainstorming: to *rapidly* generate as *many ideas* as possible. The key is not to get stuck on any one idea, especially the first one. Explain to the students that while it is quite possible that the first idea turns out to be the best solution, you do not want to become committed to that idea before other ideas, which may be better, have been considered. End the brainstorming session when a few minutes have passed with no new ideas.
- Discuss why some of the wild, impractical, or unrealistic ideas listed should not be discarded during brainstorming.
- After the initial small-group brainstorm session, have each group develop a poster

of brainstorming guidelines. Have the class provide feedback on each group's poster.

- After brainstorming, it will be time to narrow solutions. One way to do that is to give members three sticky notes each to put on their favorite ideas. Compare ideas that get the most votes with the criteria and constraints of a good solution to see which are most likely to solve the problem. (See Probe 29, "Picking the Best Solution," on p. 181 for a systematic method of doing that.)
- If student teams have difficulty coming up with brainstorming guidelines, you can suggest the following:
    - Before brainstorming, the group should discuss the problem statement until everyone is clear on the criteria and constraints for good solutions.
    - Start by spending a few silent minutes so each person can write down their own ideas.
    - Take turns, sharing one idea at a time.
    - Write down all of the suggested ideas on the board or a large sheet of paper, so everyone can see them.
    - Aim for quantity—capture as many different ideas as you can.
    - Refrain from critiquing any idea until the brainstorming session is over.
    - Welcome all ideas, no matter how crazy they may seem.
    - Build on each other's ideas.
    - Stay focused on the problem.

References

Crismond, D. P., and R. S. Adams. 2012. The informed design teaching and learning matrix. *Journal of Engineering Education* 101 (4): 738–797.

McCormick, R., P. Murphy, and M. Davidson. 1994. Design and technology as revelation and ritual. Paper presented at IDATER 1994 Conference, Loughborough, England: Loughborough University, 38–42.

National Research Council (NRC). 2012. *A framework for K–12 science education: Practices, crosscutting concepts, and core ideas.* Washington, DC: National Academies Press.

NGSS Lead States. 2013. *Next Generation Science Standards: For states, by states.* Washington, DC: National Academies Press. *www.nextgenscience.org/next-generation-science-standards.*

Engineering and Nature

Two beekeepers were talking about connections between engineering and nature. They each had a different idea:

Maia: I think engineers sometimes use designs from nature to brainstorm ways to solve human problems.

Sunny: I disagree. Engineering is about the human-designed world, not the natural world.

Who do you agree with more? ________________ Explain why you agree.

____

____

____

____

____

Ingeniería y Naturaleza

Dos apicultores estaban hablando de conexiones entre ingeniería y naturaleza. Cada uno tenía una idea diferente:

Maia: Creo que los ingenieros a veces usan diseños de la naturaleza para pensar en formas de resolver problemas humanos.

Sunny: Estoy en desacuerdo. La ingeniería se trata del mundo diseñado por los humanos, no del mundo natural.

¿Con quién estás más de acuerdo? ________________ Explica por qué estás de acuerdo.

____

____

____

____

____

Engineering and Nature

Teacher Notes

Purpose

The purpose of this assessment probe is to elicit students' ideas about how engineers use designs from nature. The probe is designed to see if students recognize that designs from nature provide creative ideas for solving problems.

Type of Probe

Opposing views

Related Key Ideas

- Nature can provide inspiration and ideas for solutions.

Explanation

The best idea is Maia's: "I think engineers sometimes use designs from nature to brainstorm ways to solve human problems." While Sunny is correct that engineering is about the human-designed world, creative ideas to solving problems can come from natural structures and processes. When brainstorming solutions, one source of ideas is to think about how nature has solved similar problems in the past. In this scenario, natural structures, such as the honeycomb —made of six-sided compartments—built by bees, is a strong and efficient way of storing things (in this case, honey). Engineers can use the honeycomb structure to design a light and strong building material.

Administering the Probe

This probe is best used with students in grades 3–12. After discussing students' answers, the probe can be extended by showing students a picture of a bee's honeycomb and asking how they might use a structure like this as inspiration to solve a problem or meet a need at home or at school. Have students sketch their ideas and briefly explain them.

Connections to the Three Dimensions (NRC 2012; NGSS Lead States 2013)

- DCI: ETS1.B. Developing Possible Solutions
- CCC: Structure and Function
- CCC: Patterns

Related Research

- Conducting workshops with university engineering students, Boga-Akyol and

Timur-Ogut (2016) found that students were initially reluctant to use natural structures to help them develop solutions, and that when they did so they tended to imitate the natural structures quite literally rather than use them as a source of ideas. However, when given a short briefing on how to use natural structures as inspirations when thinking about solutions to problems, they were more successful at arriving at creative ideas.

- Fu et al. (2014) summarized the results of hundreds of research studies on the ways that engineers use analogies, including biological structures, to solve problems, and how to support students in using these methods. Although intended for postsecondary audiences, the article provides many ideas and resources useful for K–12 teachers, such as the AskNature.org collection of resources.

Suggestions for Instruction and Assessment

- Introduce bio-inspired design (also referred to as biomimetics) as an approach to solving problems that takes inspiration from the natural world. Bio-inspired design has been used by engineers, architects, industrial designers, and artists with great success. Since the structures, materials, and systems we see in the natural world have evolved over millions of years, they can be very effective at what they do.
- This probe can be an introduction to activities in which your students learn about bio-inspired design and also to have fun with creativity. Start by showing examples of how bio-inspired design has been used by engineers. One of the most interesting is the invention of Velcro.
- Have students use the internet to determine the difference between the terms *bio-inspired design, biomimetics,* and *biomimicry.* In brief, the terms *bio-inspired design* and *biomimetics* refer to the use of natural forms to inspire human engineering, whereas *biomimicry* is the imitation of natural forms, either by humans to create engineered materials and structures, or by organisms to protect themselves. For example, a harmless red milk snake has similar coloring to a deadly coral snake. Some butterflies have coloring on their wings that look like large eyes. Ask students to think about how animals may have evolved such adaptations. (An occasional mutation may have produced a difference in coloring that allowed that animal to escape being eaten; because the animal survived, it could pass on that trait to its young.)
- Have students search the internet for other examples of "bio-inspired design," biomimetics," and "biomimicry." AskNature.org has an especially rich collection of ideas.
- Provide teams of students with a problem and examples of natural materials and structures that could help them arrive at different solutions. If students are stuck, you may follow the suggestion from Boga-Akyol and Timur-Ogut's 2016 study (discussed above) and recommend that they use the examples from nature as inspirations, not necessarily literal structures and materials to use in solving the problem.
- Discuss how scientific knowledge of biology helps engineers solve problems.

References

Boga-Akyol, M., and S. Timur-Ogut. 2016. Exploring biomimicry in the students' design process. *Design and Technology Education: An International Journal* 21 (1): 21–31.

Fu, K., D. Moreno, M. Yang, and K. L. Wood. 2014. Bio-inspired design: An overview investigating open questions from the broader field of design-by-analogy. *Journal of Mechanical Design* 136 (11): 111102-1–111102-18.

National Research Council (NRC). 2012. *A framework for K–12 science education: Practices, crosscutting concepts, and core ideas.* Washington, DC: National Academies Press.

NGSS Lead States. 2013. *Next Generation Science Standards: For states, by states.* Washington, DC: National Academies Press. *www.nextgenscience.org/next-generation-science-standards.*

Is It Affordable?

Two students were looking at an electric bicycle in a store. The label says the bike is made of titanium, a very strong and light metal. The bike has a seat made of leather, and it has a battery that will power the bike for 10 hours. The students liked the features of the bike, but it was too expensive. They each had ideas about redesigning the bike to make it more affordable:

Chenoa: If I were to make the bike more affordable, I would remove some of the more expensive features, such as the long-lasting battery. It would still be useful if it could go for just five hours.

Jerome: To make it more affordable, I don't think you would have to change the features at all. Just make it with cheaper materials and charge less.

Do you agree with Chenoa, Jerome, or both students? ________________ Explain your thinking.

____

____

____

____

____

¿Es Asequible?

Dos estudiantes miraban una exhibición de una bicicleta eléctrica. La etiqueta dice que la bicicleta está hecha de titanio, un metal muy fuerte y ligero. La bicicleta tiene un asiento que está hecho de cuero y tiene una batería que va adar potencia la bicicleta por 10 horas. Los estudiantes les gustaron las características de la bicicleta, pero era demasiado cara. Cada uno tenía ideas sobre el rediseño de la bicicleta para que sea más asequible:

Chenoa: Si tuviera que hacer que la bicicleta fuera más asequible, eliminaría algunas de las características más caras, como la batería de larga duración. Todavía sería útil si pudiera durar solo cinco horas.

Jerome: Para hacerlo más asequible, no creo que tenga que cambiar las funciones en absoluto. Solo hazlo con materiales más baratos y cobrar menos.

¿Estás de acuerdo con Chenoa, Jerome, o ambos estudiantes?__________________
Explica por qué estás de acuerdo.

Is It Affordable?

Teacher Notes

Purpose

The purpose of this assessment probe is to elicit students' ideas about trade-offs. The probe is designed to find out whether students recognize that trade-offs involve sacrificing one or more things in favor of another.

Type of Probe

Opposing views

Related Key Ideas

- Making a solution more affordable requires trade-offs between cost and other features.
- Trade-offs are decisions that involve reducing some desirable features in favor of others.

Explanation

The best answer is that either or both students have identified acceptable trade-offs for affordability. One of the first things to consider when designing a solution to a problem is how to make it affordable. If people can't buy it, there is no point in designing and manufacturing it. Designing for affordability involves finding out what people want and need most in a product, and then thinking of ways to provide that at the least cost. The clues in the probe suggest that cost could be reduced by using a less expensive battery or making the bike from less expensive materials, such as aluminum or steel rather than titanium for the frame, and plastic rather than leather for the seat. Students may come up with other ideas as well. Each of these are trade-offs in which a feature of the bike is given up in favor of affordability. While the effect of a less expensive battery may be obvious (fewer hours of travel between charges), less expensive materials also will affect the weight of the bike, its durability, and its appearance.

Administering the Probe

This probe is best used with students in grades 5–12. The probe can be used as an introduction to trade-offs when designing products that take into consideration criteria or constraints such as affordability. Extend the probe by asking students to list the things they could change to make the electric bike more affordable and how each change would affect its features. When students realize that anything they might do to reduce the cost of the bike would reduce some aspect of its performance or attraction to buyers, introduce the concept of *trade-off*—giving up some desirable characteristics of a design in favor of others (in this case, price).

Connections to the Three Dimensions (NRC 2012; NGSS Lead States 2013)

- DCI: ETS1.B. Developing Possible Solutions
- DCI: ETS1.C. Optimizing the Design Solution

Related Research

- Crismond and Adams' (2012) review of several hundred studies of how students solve problems found that "beginning designers ... focus only on positive or negative aspects of their design ideas without thinking of associated benefits and trade-offs. Informed designers balance systems of benefits and trade-offs when they consider various plans, make design decisions, and justify them ... Beginning designers can be oblivious to the unavoidable tensions and trade-offs associated with design ... Informed designers, on the other hand, are practiced at weighing and articulating both the positive features and drawbacks of ideas that they are about to select or reject and look for potential downsides even with the most promising ideas" (p. 761).

Suggestions for Instruction and Assessment

- After discussing the probe, introduce the concept of trade-off. Discuss what the trade-offs are for each change made to the design of the electric bike to make it more affordable. Ask students to think about which trade-offs they think would be most acceptable to people who might purchase the bike if the price were lower.
- A number of online websites can help students compare different brands of the same kind of product. Students can report on how different product design decisions lead to different features that are attractive to consumers as well as different prices. For example, computers with less memory or smaller screens tend to cost less. When devising an effective product comparison activity for your students, it will be important to choose a type of product that has a range of different choices available at various prices.
- Select a common product that students are familiar with, such as a smartphone, backpack, household appliance, or running shoe. Challenge them to come up with examples of features that they think people would be willing to pay a little more for, as well as other features that do not justify the added cost.
- Ask your students to name some products that many people want but few can afford. Choose one of those products and ask your students to brainstorm ideas for how it could be redesigned so that it is affordable to most people. What do they think the acceptable trade-offs would be? In other words, which features could be minimized or given up entirely so that people would still want to buy it?

References

Crismond, D. P., and R. S. Adams. 2012. The informed design teaching and learning matrix. *Journal of Engineering Education* 101 (4): 738–797.

National Research Council (NRC). 2012. *A framework for K–12 science education: Practices, crosscutting concepts, and core ideas.* Washington, DC: National Academies Press.

NGSS Lead States. 2013. *Next Generation Science Standards: For states, by states.* Washington, DC: National Academies Press. *www.nextgenscience.org/next-generation-science-standards.*

What Is a Product's Life Cycle?

A new milk company is deciding whether to use paper or plastic packaging for its milk cartons. Both cost about the same. The company's design staff has decided to look at the life cycle of each option and choose the one that is the least harmful to the environment.

Fill in the life cycle analysis chart with the words *more harm, less harm,* or *same* to compare whether paper or plastic does the greater harm to the environment. Use the chart to decide which container is the best choice.

	Gathering Raw Materials	Manufacturing the Containers	Transportation to Market	Use in the Home	Disposal After Use
Plastic Jug					
Paper Carton					

Circle the best choice: Plastic Container Paper Container

Explain how you decided which container to use.

____

____

____

____

____

¿Qué Es el Ciclo de Vida de un Producto?

Una nueva compañía de leche está decidiendo si empacar su leche en cartones de leche de papel o jarras de leche de plástico. Ambos cuestan casi lo mismo. El personal de diseño de la compañía ha decidido considerar el ciclo de vida de cada opción y elegir la que tenga el menor impacto en el medio ambiente.

Completa el cuadro de análisis del ciclo de vida con las palabras *más alto, más bajo,* o *mismo* para comparar si el papel o el plástico tienen un mayor impacto negativo en el medio ambiente. Usa la tabla para decidir qué contenedor es la mejor opción.

	Recolectando materias primas	Fabricando los Contenedores	Transporte al Mercado	Uso en el Hogar	Eliminación Después del Uso
Jarra de Plástico					
Cartón de Papel					

Encierra en un círculo la mejor opción: Jarra de Plástico Cartón de Papel

Explica cómo decidiste cuál contenedor usar.

____

____

____

____

____

What Is a Product's Life Cycle?

Teacher Notes

Purpose

The purpose of this assessment probe is to elicit students' ideas about the life cycle of a product. The probe is designed to see how students apply the concept of life cycle analysis to determine the impact of a product on the environment.

Type of Probe

Comparison chart

Related Key Idea

- Life cycle analysis includes environmental impacts of all stages of production from raw material to disposal.

Explanation

There is no single best answer to this probe. The probe provides a springboard for discussion and insight into students' thinking. Here are some ways to think about the task.

- **Gathering Raw Materials.** Plastic is made from petroleum, which is a non-renewable resource pumped from the ground. Paper is made from trees, which can be regrown if young trees are planted when fully grown trees are cut down.
- **Manufacturing.** Both paper and plastic manufacturing require energy and produce waste and air pollution. The extent of the impact depends more on the type of manufacturing plant than on the choice between the two options.
- **Transportation to Market.** "Same" is the best response for this one. Since both involve the same mode of transportation, the impact would be about the same.
- **Use in the Home.** Students may have different experiences, since there are various uses for used paper and plastic bags, but in most cases "same" is probably the best answer.
- **Disposal After Use.** In some cities, both paper and plastic can be recycled. However, there are many uses for plastic jugs, including growing plants in science classes. If they end up in landfills, paper will break down much faster than plastic. Plastic that winds up in the ocean is detrimental to marine life and beaches. So, for this part of the life cycle, plastic has a higher negative impact than paper.

Administering the Probe

This probe is best used with students in grades 3–12. Go over the life cycle analysis chart with students to make sure they understand each part of the life cycle, starting with extracting the raw materials to make the container and ending with what happens to the container after it is empty. Make sure they understand that the purpose of the rating is to compare the two products. The rating is saying which product has the higher or lower negative impact on the environment or if the products have the same impact.

Connections to the Three Dimensions (NRC 2012; NGSS Lead States 2013)

- DCI: ETS1.B. Developing Possible Solutions
- DCI: ETS2.B. Influence of Science, Engineering, and Technology on Society and the Natural World
- CCC: Cause and Effect

Related Research

- According to Crismond and Adams' (2012) review of several hundred studies of how students solve problems, "design values permeate and impact designing; they influence both how designers initially perceive and frame the task and how they evaluate ideas and complete their projects. These values can address issues related to product quality, including designing for reliability, manufacture and assembly, and designing for sustainability" (p. 763).

Suggestions for Instruction and Assessment

- Illustrate that every product, no matter how small and simple, has a life cycle and affects the environment in some way by having students list all of the objects that they see around them. Then choose one of the objects and lead a discussion about the life cycle of that object, using the chart from the probe. Assign teams of two or three students to do a life cycle analysis of other objects on the list, and then share their findings for discussion.
- Ask students to name some products that have had a negative impact on the environment, and discuss what part of their life cycle had the greatest impact. For example, many plastic objects end up in the ocean, where they cause harm to marine life. In that case, it is during the last part of its life cycle that plastic may have the greatest impact. The extraction of coal greatly affects land where it is mined, especially if strip mining is used, which can destroy entire hillsides. So coal has its greatest negative impact during the first part of its life cycle. Oil often causes the greatest impact while being transported, when accidents can occur causing oil spills at sea, or fires when oil trains are derailed. Both oil and coal cause air pollution while they are being used to generate electricity.
- For an additional challenge, ask students to research products whose negative environmental impacts were not obvious when the products were first invented. For example, when coal and oil were first introduced as a means for heating, they were considered to have far less of an impact on air quality than burning of wood and kerosene for that purpose. The use of Freon gas in refrigerators and air conditioners replaced dangerous sulfur dioxide gas. The impact of Freon on ozone in the atmosphere, allowing harmful UV rays to reach the surface and cause skin cancer, was only discovered many decades later.
- Emphasize that the life cycle of a product needs to be considered early in the design process.

References

Crismond, D. P., and R. S. Adams. 2012. The informed design teaching and learning matrix. *Journal of Engineering Education* 101 (4): 738–797.

National Research Council (NRC). 2012. *A framework for K–12 science education: Practices, crosscutting concepts, and core ideas.* Washington, DC: National Academies Press.

NGSS Lead States. 2013. *Next Generation Science Standards: For states, by states.* Washington, DC: National Academies Press. *www.nextgenscience.org/next-generation-science-standards.*

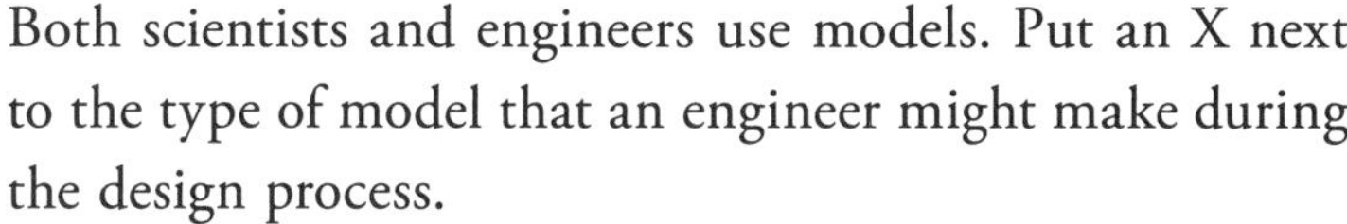

Engineers' Models

Both scientists and engineers use models. Put an X next to the type of model that an engineer might make during the design process.

___ **A.** Hand sketch

___ **B.** Block diagram

___ **C.** Computer simulation

___ **D.** Technical drawing

___ **E.** Clay model

___ **F.** Scale model

___ **G.** Physical prototype

___ **H.** Mathematical model

Explain your thinking. How did you decide if something is a model that an engineer might use?

____

____

____

____

____

Modelos Utilizados por Ingenieros

Tanto los científicos como los ingenieros usan modelos. Ponga una X al lado del tipo de modelo que un ingeniero podría usar durante el proceso de diseño.

___ **A.** Boceto hecho a mano

___ **B.** Diagrama de bloques

___ **C.** Simulación por computadora

___ **D.** Dibujo técnico

___ **E.** Modelo de arcilla

___ **F.** Modelo a escala

___ **G.** Prototipo físico

___ **H.** Modelo matemático

Explica lo que piensas. ¿Cómo decidiste si algo se considera un modelo que un ingeniero podría usar?

____

____

____

____

____

Engineers' Models

Teacher Notes

Purpose

The purpose of this assessment probe is to elicit students' ideas about different types of models commonly developed by engineers. The probe is designed to reveal whether students recognize that just like scientists, engineers make and use a variety of model types, including physical, conceptual, and mathematical models.

Type of Probe

Justified list

Related Key Idea

- Engineers make and use various kinds of models at different stages of a design process.

Explanation

The best answer is that all of the things on the list are models made by engineers. Hand sketches, block diagrams, technical drawings, and scale models are examples of conceptual models. Mathematical models are another type of model which also includes computer simulations. The clay model and the physical prototype are different kinds of physical models. Being able to select and develop various models is a critical engineering design skill.

Administering the Probe

This probe is best used with students in grades 3–12. For younger students, we suggest simplifying the list of models to those likely to be very familiar to students in the class. These would include obvious items such as sketches, drawings, and clay models, but also consider that a computer model can be related to the familiar idea of a video game that plays on a computer or a phone. Explain to students that a video game is a type of computer model and that engineers can use such a visual model to describe a product design. If after the introductory explanations, students are still not familiar with any of the models on the list, have them put a question mark next to it or describe it in terms the students understand. The probe can be extended by having the students describe each type of model they selected and how they think it is used in an engineering design process.

Connections to the Three Dimensions (NRC 2012; NGSS Lead States 2013)

- DCI: ETS1.B. Developing Possible Solutions
- DCI: ETS1.C. Optimizing the Design Solution
- SEP: Developing and Using Models.
- CCC: Systems and System Models

Related Research

- Crismond and Adams' (2012) review of several hundred studies of how students solve problems noted that "modeling can involve building a physical prototype—an approximation of the product along one or more dimensions of interest, using easy-to-fabricate modeling materials, like cardboard and duct tape, or easy-to-assemble structural elements, like LEGOs™. Mathematical models, including those that are the basis of computer simulations, can represent the problem or potential solutions and act as cognitive devices to enable thinking. These approaches can help students visualize their product ideas more easily, especially students with modest drawing skills" (p. 759).
- Zubrowski (2002) summarized the results of several years of research and development by recommending a specific use of models for teaching engineering design. The approach starts with a challenge "to design and build a working model of a technological artifact, such as a flying toy, windmill, water wheel, or balloon-powered car, with a limited set of materials and initial performance criteria. After preliminary models are designed and tested, there is a shift to a standard model, which is used to carry out inquiry providing for a more controlled context for introducing basic science concepts. … After some experimenting, there is a return to the preliminary design models that are either modified based on the newly gained knowledge or completely redeveloped and tested" (p. 49).

Suggestions for Instruction and Assessment

- To help develop their ability to use models effectively, students should learn about different kinds of models and their purposes. This probe can be extended by using it as a card sort, using the "Model Deck" and "Description Deck" information that follows. Print each type of model on a card. On another set of cards, print the descriptions of the models listed in the "Description Deck" section that follows. (Print the descriptions without the numbers. The numbers are provided here for you to match the description to the type of model.) Have students work in small groups to discuss and match the model type with its description.

Model Deck

1. Hand Sketch
2. Block Diagram
3. Computer Simulation
4. Technical Drawing
5. Clay Model
6. Scale Model
7. Physical Prototype
8. Mathematical Model

Description Deck

1. Visual image of a design drawn quickly to describe key features of an idea
2. Drawing using words, shapes, and lines to illustrate a system or process
3. A programmed representation showing how a design would behave under different conditions
4. A detailed drawing with precise dimensions to communicate how something functions or is to be constructed

5. Physical model made out of soft material to let you see the shape of a designed object
6. Physical model made of various materials (cardboard, wood, plastic, metal) that has the same proportions of a design object, but is usually smaller in size
7. Physical model that allows a design to be tested to see if it solves the problem
8. An abstraction using numbers and symbols to describe a designed object or system

- Extend the card sort activity by having students find examples of each type of model.
- Be sure to discuss with students that engineers not only use models, but also make them, and that it is in the making of a model that an engineer or scientist gains deeper understanding of the problem and the design.
- Point out different uses of models for engineering: to visualize one or more solutions in order to understand and compare their key properties; to help choose which solution is best; and to try out different versions of a design on a physical or computer model to save time and money that would be needed to build several full-scale models. Ask students if they can think of other uses of models for engineering.
- Have students compare the types of models scientists use with models used by engineers. Which of the models on the list are used by scientists, engineers, or both?
- Crismond and Adams (2012) comment that "having beginning designers fabricate flawed yet working models provides them with an initial taste of success, yet leaves plenty of room for them to improve their devices over multiple iterations" (p. 759).
- The instructional sequence described in Zubrowski's standard model (2002) has students first design an initial prototype without suggestions or help from the instructor, and then build a teacher-supplied model of the same device. After testing both models, students attempt to synthesize ideas to create an optimal solution.

References

Crismond, D. P., and R. S. Adams. 2012. The informed design teaching and learning matrix. *Journal of Engineering Education* 101 (4): 738–797.

National Research Council (NRC). 2012. *A framework for K–12 science education: Practices, crosscutting concepts, and core ideas.* Washington, DC: National Academies Press.

NGSS Lead States. 2013. *Next Generation Science Standards: For states, by states.* Washington, DC: National Academies Press. *www.nextgenscience.org/next-generation-science-standards.*

Zubrowski, B. 2002. Integrating science into design technology projects: Using a standard model in the design process. *Journal of Technology Education* 13 (2): 48–67.

Picking the Best Solution

Criteria	Weight	Pizza Pal	Sal's Pizza	Pizza Roma
Size of tables	5	3	5	3
Good pizza	4	4	4	3
Not too loud	3	3	1	2
Total		10	10	8

A small group of friends want to have a pizza party. They can't decide whether to go to Pizza Pal, Sal's Pizza, or Pizza Roma. Instead of arguing, they think like engineers and list their criteria. They want to sit around one table, they want good pizza, and they want to hear each other talk.

To create the decision matrix as shown above, the friends list their criteria on the left and assign a weight (a numerical score) to each one. The most important criterion is to have room for them to all sit comfortably, so size of table gets the most weight. For each of the possible solutions, they assign points up to the maximum weight. Then they add up the numbers to see which solution seems best, based on all their criteria.

Using the matrix, which pizza place would you choose? ____________________ Explain why that would be the best choice.

____

____

____

____

____

____

Elegir la Mejor Solución

Criterios	Puntuación	Pizza Pal	Sal's Pizza	Pizza Roma
Mesas Grandes	5	3	5	3
Buena Pizza	4	4	4	3
No Muy Ruidoso	3	3	1	2
Total		10	10	8

Un pequeño grupo de amigos quiere tener una fiesta de pizza. No pueden decidir si ir a Pizza Pal, Sal's Pizza, o Pizza Roma. En lugar de discutir, piensan como ingenieros y enumeran sus criterios. Quieren sentarse alrededor de una mesa, quieren buena pizza, y quieren escucharse mutuamente.

Para crear la matriz de decisión como se muestra arriba, los amigos enumeran sus criterios a la izquierda y asignan una puntuación numérica a cada uno. El criterio más importante es tener espacio para que todos se sienten juntos, entonces las mesas grandes obtenien el puntaje más alto. Para cada una de las soluciones posibles, asignar puntos hasta el peso máxima. Luego suma los números para ver qué solución parece mejor, basado en todos los criterios.

Usando la matriz, ¿qué pizzería elegirías? ____________________ Explica por qué esa sería la mejor opción.

____

____

____

____

____

____

Picking the Best Solution

Teacher Notes

Criteria	Weight	Pizza Pal	Sal's Pizza	Pizza Roma
Size of tables	5	3	5	3
Good pizza	4	4	4	3
Not too loud	3	3	1	2
Total		10	10	8

Purpose

The purpose of this assessment probe is to provide practice in using a decision matrix. The probe is designed to see if students are able to interpret a decision matrix and to help you determine difficulties they may encounter.

Type of Probe

- Comparison chart

Related Key Ideas

- A decision matrix is one way to systematically compare solutions to see which is best.
- Trade-offs are decisions that involve reducing some desirable features in favor of others.

Explanation

The best answer is "Sal's Pizza." Look for evidence in the students' responses that they have paid attention to what the scores mean. Even though Pizza Pal and Sal's Pizza have the same total score, students need to consider the weight of the criteria—space for everyone to sit at the table is the most important. An indication that a student understands this would be the response, "Sal's is the only pizza place that has tables big enough for the group to all sit around." A student could also add that although it's a little noisy at Sal's, the pizza is at least as good as Pizza Pal and better than Pizza Roma.

Administering the Probe

This probe is best used with students in grades 6–12. The probe is intended to be used to introduce students to a process they can use after brainstorming. Explain that after they have generated several solutions and narrowed down the three or four most promising ones, engineers often use a systematic method called a *decision matrix* to choose the best solution before they build and test a prototype. The pizza decision matrix is used as a familiar example to see how a decision matrix works. Extend the probe by asking students to explain how a decision matrix approach can be used to decide which engineering solution is best for developing and testing.

Connections to the Three Dimensions (NRC 2012; NGSS Lead States 2013)

- DCI: ETS1.B. Developing Possible Solutions
- DCI: ETS1.C. Optimizing the Design Solution
- SEP: Analyzing and Interpreting Data

Related Research

- Crismond and Adams' (2012) review of several hundred studies of how students solve problems found that "beginning designers ignore or pay too little attention to design criteria and constraints, and focus only on positive or negative aspects of their design ideas without thinking of associated benefits and trade-offs. Informed designers balance systems of benefits and trade-offs when they consider various plans, make design decisions, and justify them" (p. 761).

Suggestions for Instruction and Assessment

- The use of a decision matrix is a good time to introduce (or reinforce) the idea of trade-offs. In this case, Sal's is a better choice overall, but it is louder than the other options. So, it is a trade-off—room for everyone, good pizza and service, but harder to hear each other.
- Have students notice that a decision matrix only includes criteria. Constraints are typically not involved because if one constraint is violated the solution is rejected.
- Form teams of two or three students to practice creating and using decision matrix charts. List a few problems (preferably with local interest) from which students could choose, for which two or three alternative solutions have been proposed. Have students first list their criteria for making a choice, prioritize the criteria, and then make a decision matrix to compare the alternative solutions. For example, one problem could be that the cafeteria is too crowded. Criteria could include more comfortable eating conditions, less time spent in the serving line, and a lower noise level. Alternative solutions could include having more lunch periods or expanding the cafeteria by reducing the size of the gym.
- Discuss the value of having a decision matrix with students, but be sure to note that there can be intangible decision factors beyond the main ones in the matrix and that an important part of talking with clients is understanding what all of the factors are that concern them.
- Have your students search the web to learn more about a decision matrix—also called a Pugh chart, named for its inventor, Stuart Pugh. Students can research Pugh's life and find out about the broader system of ideas, called "Total Design," that he created to guide any engineering project, regardless of topic or field.
- Have students look up Pugh charts on the internet to learn about different ways of scoring them.

References

Crismond, D. P., and R. S. Adams. 2012. The informed design teaching and learning matrix. *Journal of Engineering Education* 101 (4): 738–797.

National Research Council (NRC). 2012. *A framework for K–12 science education: Practices, crosscutting concepts, and core ideas.* Washington, DC: National Academies Press.

NGSS Lead States. 2013. *Next Generation Science Standards: For states, by states.* Washington, DC: National Academies Press. *www.nextgenscience.org/next-generation-science-standards.*

Designing With Math and Science

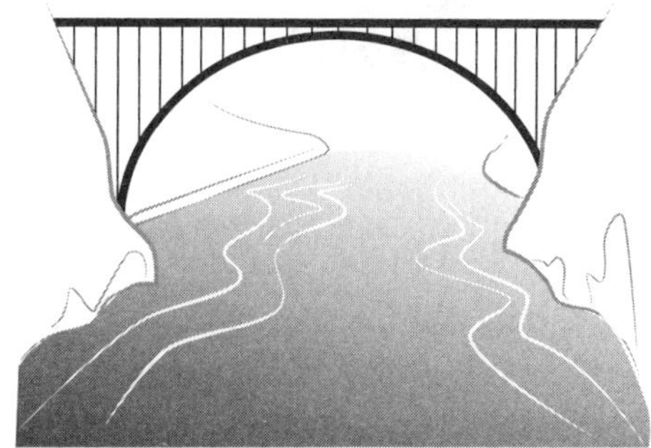

Civil engineers learn math and science so that they can design structures, such as bridges, that can handle heavy use for many years and resist the forces of nature. Civil engineers have to answer many questions while designing a bridge. Some of those questions are listed below.

Check the boxes in the table to indicate if the engineers who designed this bridge needed to apply their knowledge of math, science, or both to answer each question.

Question	Math	Science	Both
1. How can the steel of the bridge be protected so it does not rust and corrode?			
2. For the bridge to be 100 feet high, what diameter circle should be used to design the arch?			
3. What is the strongest wind this bridge has to resist?			
4. What is the maximum weight that the concrete foundation under the bridge will need to support?			

Explain your thinking.

Diseñando con Matemática y Ciencia

Los ingenieros civiles aprenden matemáticas y ciencias para poder diseñar estructuras, como puentes, que puedan manejar el uso intensivo durante muchos años y resistir las fuerzas de la naturaleza. Hay varias preguntas que los ingenieros civiles deben responder al diseñar un puente.

Marque las casillas para indicar si los ingenieros que diseñaron este puente necesitan aplicar sus conocimientos de matemática, ciencia, o ambos.

Pregunta	Matemática	Ciencia	Ambos
1. ¿Cómo se puede proteger el acero del puente para que no se oxida y corroe?			
2. Para que el puente tenga 100 pies de altura, ¿qué círculo de diámetro se debe usar para diseñar el arco?			
3. ¿Cuál es el viento más fuerte que este puente tiene que resistir?			
4. ¿Cuál es el peso máximo que la base de concreto debajo del puente deberá soportar?			

Explica lo que piensas.

___

___

___

___

___

Designing With Math and Science

Teacher Notes

Purpose

The purpose of this assessment probe is to elicit students' ideas about how science and math are both used in engineering. The probe is designed to get students to think about the various ways science and math are applied to designing a physical structure such as a bridge.

Type of Probe

Justified list

Related Key Idea

- Most engineering problems require the application of math and science.

Explanation

The best answer is "both" for ALL of the questions—almost all engineering needs both math and science!

- The first one needs science to know about chemistry of corrosion, and math to calculate how much coating thickness is needed for protecting steel for a certain number of years.
- The second one involves knowledge of science (physics and chemistry) for the material properties of steel (strength, flexibility), and geometric math to calculate the dimensions of the arch shape.
- The third one requires the science of weather (meteorology) to know the likely wind conditions in the region, the sciences of physics and chemistry for the material properties of steel (strength, flexibility), and math to calculate the wind resistance capacity of the bridge for its given dimensions.
- The fourth one involves knowledge of science (physics and chemistry) for the materials properties (strength, durability) of concrete, and math to calculate the required weight-bearing capacity of the foundation.

This engineering problem is not unique—most require science and math in a clearly relevant way.

Administering the Probe

This probe is best used with students in grades 6–12. Encourage students to support their answers with specific examples of the knowledge needed in science or math for each question.

Connections to the Three Dimensions (NRC 2012; NGSS Lead States 2013)

- DCI: ETS2.A. Interdependence of Science, Engineering, and Society
- SEP: Using Mathematics and Computational Thinking

Related Research

As shown in the following two studies, students' positive attitudes toward science, math, and engineering can be affected by engineering programs at school, and by enlisting the help of parents:

- Hirsch et al. (2007) administered surveys to 890 fifth through eighth grade students to determine differential effects of various pre-engineering curricula. Approximately half of the students were exposed to pre-engineering concepts in various ways in their science and math classes. The researchers found that students who had received pre-engineering instruction were more confident in their science, math, and engineering capabilities; had greater knowledge of types of engineers; and were better able to explain what different kinds of work engineers do.
- Rozek et al. (2017) evaluated the long-term effects of an intervention designed to help parents convey the importance of math and science courses to their high-school–aged children. The intervention consisted of a website, two brochures, and advice for parents about how to communicate the information to their children. The researchers found that the intervention improved math and science standardized test scores on a college preparatory examination (ACT) by 12 percentile points. Greater high-school STEM preparation (STEM course-taking and ACT scores) was associated with increased STEM career pursuit (STEM career interest, the number of college STEM courses, and students' attitudes toward STEM) five years after the intervention. Results suggest the intervention can affect STEM career pursuit by helping parents recognize their essential role in encouraging their children's STEM interests and willingness to take STEM courses in high school.

Suggestions for Instruction and Assessment

- Have your students think about engineering other things the world will need: For example, an affordable electric car, a longer battery-life cell phone, or an artificial heart are all great ways to show students how naturally science and math come together in engineering to create exciting and impactful solutions to help humanity. Students are often disengaged from science and math because they don't see the relevance to their life and future, and engineering provides a natural motivation for engaging students' interest in science and math learning.
- There are a variety of bridge-building challenges students can engage in using everyday materials such as craft or stirring sticks, toothpicks, newspaper, or dried pasta (these can be found with an internet search for "bridge building challenges"). Have students generate a list of the scientific and mathematical knowledge they used to design their bridge.
- Besides designing bridges, civil engineers engage in other kinds of design projects, including roads, buildings, airports, tunnels, dams, and water treatment plants. List these on the board and ask students to name some of the ways that math and science would be applied to these projects.
- Invite a civil engineer to the classroom to talk about his or her work with students.

- Provide high school students with the following additional questions that civil engineers face when designing a bridge. Have them discuss what science, mathematical, or other knowledge would be needed to answer the questions.
  - **What shape does the steel structure need to be to support the roadway, so it does not sag and bend?** This question requires an understanding of the properties of steel and other materials, as well as an understanding of the way that different geometric shapes resist forces. Answering this complex question requires knowledge of both science and math.
  - **How will traffic on the bridge increase in future years?** Answering this question involves some of the social sciences, such as economics and sociology, to predict how the population and commerce will grow. Those projections will need to be turned into mathematical calculations to see how they are likely to affect the bridge structure.
- In addition to math and science, have students consider what other disciplines are used in engineering. For example, how are literacy, psychology, geography, anthropology, history, and art used in engineering?

References

Hirsch, L. S., J. D. Carpinelli, H. Kimmel, R. Rockland, and J. Bloom. 2007. The differential effects of pre-engineering curricula on middle school students' attitudes to and knowledge of engineering careers. Paper presented at the ASEE/IEEE 37th Frontiers in Education Conference, Milwaukee. *https://ieeexplore.ieee.org/stamp/stamp.jsp?arnumber=4417918.*

National Research Council (NRC). 2012. *A framework for K–12 science education: Practices, crosscutting concepts, and core ideas.* Washington, DC: National Academies Press.

NGSS Lead States. 2013. *Next Generation Science Standards: For states, by states.* Washington, DC: National Academies Press. *www.nextgenscience.org/next-generation-science-standards.*

Rozek, C. S., R. C. Svoboda, J. M. Harackiewicz, C. S. Hulleman, and J. S. Hyde. 2017. Utility-value intervention with parents increases students' STEM preparation and career pursuit. *Proceedings of the National Academy of Sciences* 114 (5): 909–914.

Testing for Success

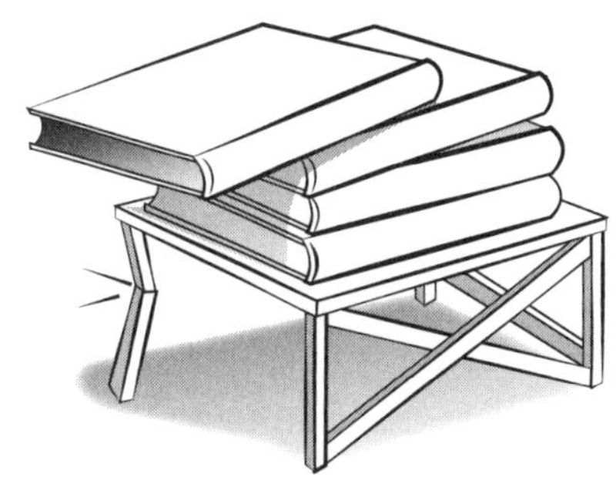

Three students are building a model house using craft sticks that each cost $100 in play money. Their challenge is to construct a framework that will support the weight of at least three books at the lowest possible cost. This is what they talked about as they tested their model:

Adam: Our model house is supposed to hold at least three books and it only starts to break when we load it with four books. Our test was successful. Time to celebrate!

Jayanti: Hey, it's great we got it to work, but take a close look. You can see it's only breaking in one place where we forgot to put in triangle braces for support. If we put braces in the place where it broke, it will cost only a little bit more but we might be able to make it a lot stronger and safer.

Ishan: Let's repair it by replacing the broken columns and improve it by putting in more triangle braces in all of the corners. Then we can test it to check the strength—maybe we can get to six books and be twice as safe as the requirement!

Who do you think has the best idea about the result of testing their model? ________________ Explain your thinking.

____

____

____

____

____

Prueba Para el Exito

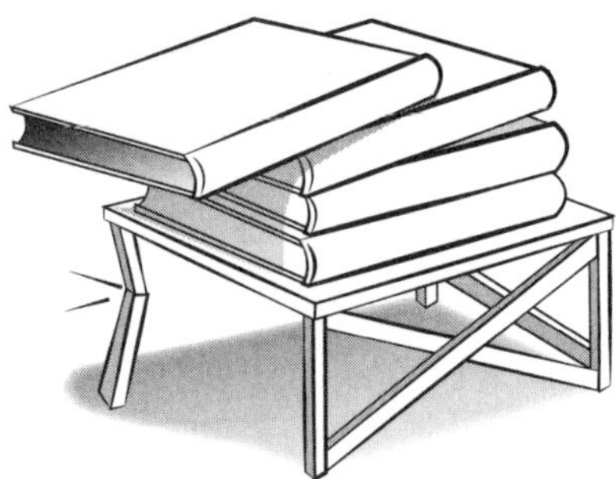

Tres estudiantes están construyendo un modelo de casa con palitos de helado que cuestan $100 en dinero de juego. Su desafío es construir un marco que soporte el peso de al menos tres libros al menor costo posible. De esto hablaron cuando probaron su modelo:

Adam: Se supone que nuestro modelo de casa tiene al menos tres libros y solo comienza a romperse cuando la cargamos con cuatro libros. Nuestra prueba fue exitosa. ¡Tiempo para celebrar!

Jayanti: Oye, es genial que esta trabajando, pero fíjate bien. Puede ver que solo se está rompiendo en un lugar donde olvidamos poner llaves triangulares para soporte. Si colocamos frenos en el lugar donde se rompió, costará solo un poco más, pero podríamos hacerlo mucho más fuerte y seguro.

Ishan: Reparémoslo reemplazando las columnas rotas y mejorándolo colocando más llaves triangulares en cada una de las esquinas. Luego podemos probarlo para verificar la resistencia, ¡tal vez podamos llegar a seis libros y estar dos veces más seguros que el requisito!

¿Quién crees que tiene la mejor idea sobre el resultado de probar su modelo? ________________ Explica lo que piensas.

____

____

____

____

____

Testing for Success

Teacher Notes

Purpose

The purpose of this assessment probe is to elicit students' initial ideas and attitudes about the procedure that should be followed when testing models to determine if a design solves the problem. The probe is designed to see if students recognize how a design can be improved based on a careful analysis of the result of the test.

Type of Probe

Friendly talk

Related Key Ideas

- Engineers make and use various kinds of models at different stages of a design process.
- Solutions must be tested to see if they meet the criteria and constraints of the problem.
- A prototype is a model that can be tested to check if the design solves the problem.

Explanation

The best idea is Jayanti's: "Hey, it's great we got it to work, but take a close look. You can see it's only breaking in one place where we forgot to put in triangle braces for support. If we put braces in the place where it broke, it will cost only a little bit more but we might be able to make it a lot stronger and safer." Jayanti has paid close attention to the results of the test. She did not simply observe if they succeeded in meeting their goal. She also noted that the model broke in a particular way. By noticing where and how the structure broke, she showed how their design could be improved with little additional cost. Ishan also had a good suggestion. By adding additional triangles in each of the corners, it could be made even safer. However, that would cost a lot more, so fewer people would be able to afford the actual house when it is built.

Administering the Probe

This probe is best used with students in grades 3–12. It can be embedded into a design task right before students test their models, to determine whether students are more focused on "winning" the challenge or thinking about the principles of good design.

Connections to the Three Dimensions (NRC 2012; NGSS Lead States 2013)

- DCI: ETS1.C. Optimizing the Design Solution
- SEP: Developing and Using Models

Related Research

The context of this probe often involves a student competition to engineer some type of load-bearing structure such as a bridge or tower. The intent of the probe is not to recommend competitive activities as a superior method for learning engineering design; however, competitive activities or behavior are common and should be used thoughtfully with the knowledge that other approaches can work equally well and possibly better with underserved or low-esteem students. The following research summaries address competitions in engineering:

- Sadler, Coyle, and Schwartz (2000) developed a number of engineering middle school units aimed at reducing the competition that typically characterizes engineering activities, such as building the tallest tower or the strongest bridge. Instead, teams of students tested their designs against nature, collaborating with their teammates and borrowing ideas from other teams during frequent tests in which the whole class participated. Data for evaluating and improving the program were collected from pilot studies with 12 teachers nationwide and their students, using pre- and post-tests (N=457), student interviews, classroom observations, and storyboard analysis. The researchers concluded that such non-competitive challenges help erase gender disparity common in engineering and physical science challenges.
- Beyer and Auster (2014) conducted a study to determine if competition made a difference when engineering challenges are presented in a museum setting. One hundred small groups of museum visitors in the age range of 8 to 14 years old were challenged to design and test model bobsleds. Half of the groups were presented with the challenge as a competition with prior teams, whose results were posted in the museum. No records were posted for the other half of the groups. No difference was observed between the two conditions, and there were no differences between boys and girls.

Suggestions for Instruction and Assessment

Testing a design to see if it meets the criteria and constraints of a problem is a critical aspect of an engineering design process. To help your students develop this skill, it is important to start with a design problem that will lead to a meaningful and satisfying testing experience. Specifically, we recommend the following guidelines suggested by Sadler, Coyle, and Schwartz (2000), and modified by our own experiences in developing engineering activities for students:

- **Start with an engaging "make" activity.** There are hundreds of simple mechanisms that students can build that capture their interest. In the past these were often referred to as "make-it-and-take-it" projects. They range from mousetrap cars to parachutes, to windmills for generating electricity. An important difference in our recommendation is to see these projects as starting points, and to involve students in the design process of improving the technology. After all, that is what engineers do most of the time—improve existing technologies.
- **Make sure the activity has a clear goal.** If it's a mousetrap car, the goal might be to travel as far as possible on one setting of the mousetrap. If it's a parachute, it might be the longest glide time. If it's a windmill for generating electricity, the goal may be

to generate the greatest electric current with a given breeze. Students need such a goal to determine if they are making progress. Also, it's important that students "buy in" to the goal of the activity and understand what is expected of them.

- **Choose a project that can be significantly improved.** Students should have an opportunity to introduce design improvements so that their design performs better than the first "make" activity. There should be more than one variable so they have a chance to be creative.
- **Allow for multiple public tests.** Student teams should be able to test their improvements and watch the tests of others. They should be encouraged to learn from and borrow ideas from other teams, since the goal is to create the most successful possible design. There is no point in reinventing the wheel.
- **Encourage feedback.** Students must be able to use the feedback that results from testing their ideas, so they can improve their designs and occasionally contrast their prior beliefs that might be challenged by unexpected results.
- **Motivate students to keep records.** Although many students tend to resist writing down the results of their efforts, helping them create engineer notebooks with numbered and dated pages can sometimes motivate record keeping. Another method is to have them use phones or tablets to make short videos of the project.
- **De-emphasize competition.** Although the jury is out on the value of competition, it should not be emphasized too much, so that students do not think primarily about beating the other teams, rather than focusing on their design process.

References

Beyer, M., and R. Auster. 2014. Assessing competition in engineering (ACE) research report. Boston: Museum of Science. *www.informalscience.org/sites/default/files/2015-06-15_2014_Assessing_Competition_in_Engineering.pdf.*

National Research Council (NRC). 2012. *A framework for K–12 science education: Practices, crosscutting concepts, and core ideas.* Washington, DC: National Academies Press.

NGSS Lead States. 2013. *Next Generation Science Standards: For states, by states.* Washington, DC: National Academies Press. *www.nextgenscience.org/next-generation-science-standards.*

Sadler, P. M., H. P. Coyle, and M. Schwartz. 2000. Engineering competitions in the middle school classroom: Key elements in developing effective design challenges. *Journal of the Learning Sciences* 9 (3): 299–327.

Making It Better

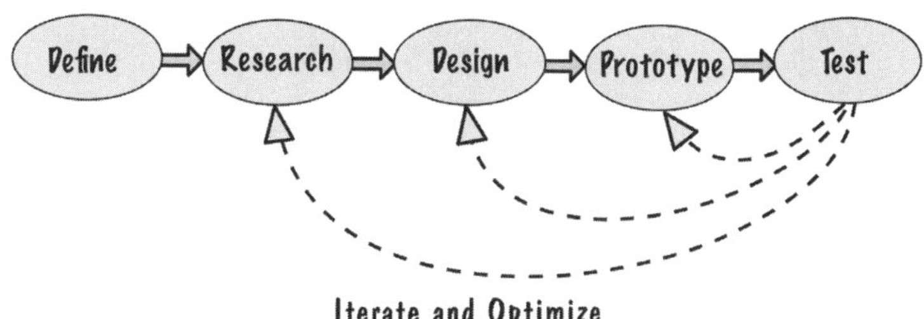

Several engineers have started their own sports products company. They are letting students from a local school test the company's new backpack design. The engineers are meeting with school students to hear what they think about the backpack prototypes. Three students are sharing their experiences and ideas. The chart above of an engineering design process is hanging on the wall.

Steven: I have a long walk home with a heavy backpack full of books, and the main zipper is tearing loose from the cloth. You need to make a new prototype with stronger stitching.

Divya: My zippers are working fine, but when the backpack is full it strains my neck and back. You should improve the design so that it reduces the strain—maybe a hip belt or something?

Alok: My backpack works well, but I have another problem. My mobile phone runs out of power! Can you add solar power technology to the backpack so it can charge my phone if I sit by a sunny window?

Circle the name of one of the students and explain how that student is following the engineering design process.

____

____

____

____

____

Haciéndolo Mejor

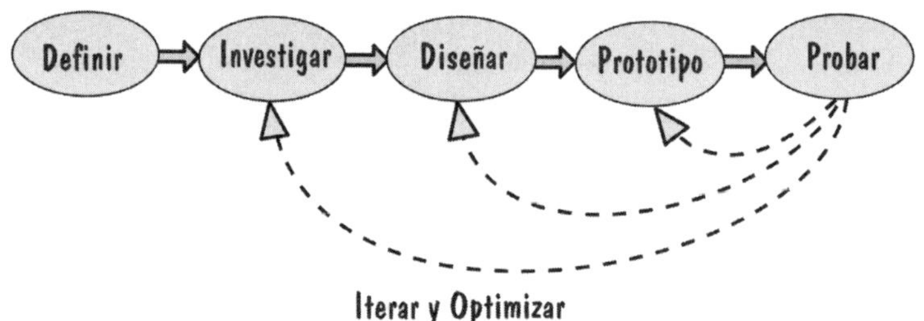

Varios ingenieros han comenzado su propia empresa de productos deportivos. Están permitiendo que los estudiantes de una escuela local prueben el nuevo diseño de la mochila de la compañía. Los ingenieros se reúnen con estudiantes de la escuela para escuchar lo que piensan acerca de los prototipos de mochilas. Tres estudiantes comparten sus experiencias e ideas. La tabla de un proceso de diseño de ingeniería está colgada en la pared.

Steven: Tengo un largo camino a casa con una mochila pesada llena de libros, y la cremallera principal se está soltando de la tela. Necesita hacer un nuevo prototipo con costuras más fuertes.

Divya: Mis cremalleras funcionan bien, pero cuando está llena, la mochila me estira el cuello y la espalda. Debería mejorar el diseño para que reduzca la tensión, ¿tal vez un cinturón de cadera o algo así?

Alok: Mi mochila funciona bien, pero tengo otro problema. ¡Mi teléfono móvil se queda sin energía! ¿Puede agregar tecnología de energía solar para que pueda cargar mi teléfono si me siento junto a una ventana soleada?

Encierra en un círculo el nombre de uno de los estudiantes y explica cómo el estudiante sigue el proceso de diseño de ingeniería.

____

____

____

____

____

Making It Better

Teacher Notes

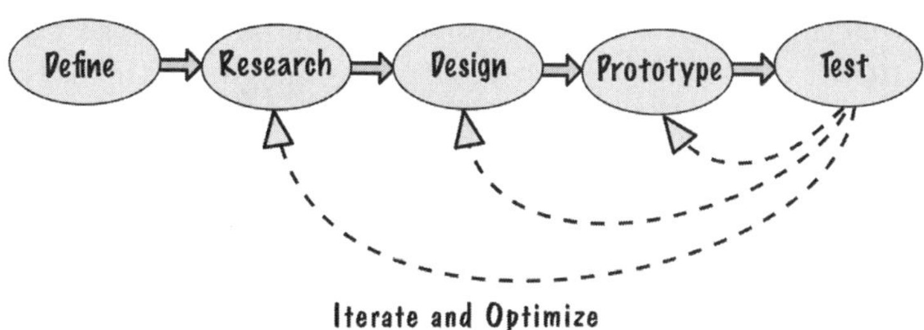

Purpose

The purpose of this assessment probe is to elicit students' ideas about different ways to optimize a design that has been successful in the past, but that still needs improvement. The probe is designed to see if students recognize that there are multiple pathways to iterating and optimizing a design.

Type of Probe

Friendly talk

Related Key Ideas

- Optimization involves further tests and improvements to find the best possible solution.
- Many different methods can be used to optimize a design.

Explanation

All three students are following an engineering design process (EDP) as illustrated on the chart. The dotted lines with arrowheads in the diagram of the EDP have an important meaning. Here's how the three students map onto that diagram:

- Steven proposes taking the shortest loop. Just make and test a new prototype.
- Divya thinks they can improve the backpack design to reduce neck and back strain. She proposes taking the middle loop back to the design phase.
- Alok doesn't have a problem that needs fixing, but as often happens with user feedback, he shares a new need—having a backpack that could charge a mobile phone or other devices. He recognizes that the engineers will have to go all the way back to the research phase to learn about solar panel technologies that might work in a backpack design, as well as further discussions with students to learn more about their needs that could be met with an improved backpack design.

This process of testing, redesigning, and testing again is called *iteration*. Continuing to iterate until the best possible design is reached is called *optimization*.

Administering the Probe

This probe is best used with students in grades 6–12. The backpack is a familiar item to students over a wide range of ages. This probe may be used with students down to third grade by first discussing the vocabulary of a simplified EDP (see Probe 17, p. 109, for details of an EDP suitable for elementary grades) as well as the basic concept of charging cell phones with solar panels. For all ages, it's good to review the EDP diagram with students first, especially if they are used to seeing a different diagram of an engineering design process. If students are not familiar with the terms *iteration* and *optimization,* you may clarify those terms for them.

Connections to the Three Dimensions (NRC 2012; NGSS Lead States 2013)

- DCI: ETS1.C. Optimizing the Design Solution
- CCC: Cause and Effect

Related Research

- Crismond and Adams' (2012) review of hundreds of studies of engineering education notes that science investigations generally aim at uncovering causal relationships. In contrast, engineering design aims to optimize a desired outcome. In the case of this probe, for example, the intent of the research proposed by Alok is not to uncover a causal relationship, but rather to learn enough so as to redesign the backpacks to meet students' needs. Such investigations address technical and social dimensions, and "tend to be more consistent with everyday problem solving" (Crismond and Adams 2012, p. 765).

Suggestions for Instruction and Assessment

- Extend the probe by encouraging your students to debate which is the best approach. As illustrated in the EDP diagram, all three approaches are valid, depending on the particular situation.
- If time permits, the probe can serve as an engaging entry into an activity in which students design their own improved backpack by sketching their design ideas on paper. The backpack problems in this scenario are all real problems (see Talbott et al. [2009] for a study of bodily strain problems from backpacks), and there are more. A simple 20-minute activity is for students to list their own problems and improvement ideas for backpacks.
- A deeper extension that can occupy a full class period is to introduce the probe as an example, and then have students work in teams to select products from their daily life that have problems needing improvement. Each team then selects a product and follows an EDP to define the problem, brainstorm solutions, and select one solution to develop further into a "concept" solution described by a paper sketch and simple description. In addition to the backpack, some examples of daily objects familiar to students are toothbrushes, refrigerators, bicycles, phones, chairs, desks, lamps, beds, and shopping bags. You might be pleasantly surprised by the ideas that can come from students' imaginations.
- The best approach when optimizing a design may depend on the product that is being re-designed. The backpack is a product that has been used by people for centuries and has undergone improvements ranging from minor changes in shape all the way to revolutionary changes in technology, such as new breathable and waterproof fabrics. It's instructive to have

students discuss other types of products to add complementary perspective to the backpack case. Ask your students which loop in the EDP diagram would be best to solve the following problems:

- A mobile phone company discovers that its latest batch of phones is developing cracks in the plastic within only weeks of use. (A new prototype with stronger plastic may fix the problem.)
- A manufacturer learns that its cat toy is only played with by about half of the cats that used it during trial tests. (The toy should be redesigned and tested again.)
- A process chemist who works for a factory discovers that the plant's old filtration system is not removing all of the chemicals in the wastewater coming from a new manufacturing process. (Further research will be needed to determine what chemicals have not been removed and which processes will be needed to remove them.)

- Have students compare and contrast the crosscutting concept of cause and effect in science with the engineering practice of optimizing a desired outcome.

References

Crismond, D. P., and R. S. Adams. 2012. The informed design teaching and learning matrix. *Journal of Engineering Education* 101 (4): 738–797.

National Research Council (NRC). 2012. *A framework for K–12 science education: Practices, crosscutting concepts, and core ideas.* Washington, DC: National Academies Press.

NGSS Lead States. 2013. *Next Generation Science Standards: For states, by states.* Washington, DC: National Academies Press. *www.nextgenscience.org/next-generation-science-standards.*

Talbott, N. R., A. Bhattacharya, K. G. Davis, R. Shukla, and L. Levin. 2009. School backpacks: It's more than just a weight problem. *Work* 34 (4): 481–494.

Index

Index

Index

Index

Index

Index

Index

Index

Index